The 7 Steps of Data Analysis:

A Manual for Conducting a Quantitative Research Study

First Edition

WILLIAM M. BANNON, JR.

StatsWhisperer Press

New York

Senior Editor: Tamara A. Bannon
Assistant Editor: Lori Bruno-Lee
Cover Design: Tamara A. Bannon

Library of Congress Cataloging-in-Publication Data

Bannon, Jr., William M.,

     The 7 Steps of Data Analysis / William M. Bannon, Jr. −1st ed.   p. cm.

          Includes bibliographical references

          ISBN 978-0-615-85729-9

          Library of Congress Control Number 2013949397

          1. Steps of Data Analysis (Statistics)—Handbooks, manuals, etc.

          I. Authorship   I. Bannon, Jr., William M.   II. Title

ISBN   978-0-615-85729-9

Printed in the United States of America

Dedication:

To My Wife Tammy and Children Reese, Autumn, Luciana, and Austin
Without Whose Support This Book Would Not Be Possible.

And My Father William and Mother Susanne
Without Whom Life Would Not Be Possible.

Acknowledgments

I would like to extend my thanks to all of the authors, educators, and colleagues who contributed their effort, talent, and good will to the learning experience that helped form the principles presented in this book. In particular, I would like to thank my original research mentor Dr. Chaya Piotrkowski for not only introducing me to the principles of quantitative and qualitative research, but also for illustrating the marvelous things possible through each. Also, I would like to thank my long time mentor Dr. Mary McKay, who worked closely with me for many years as I developed this system through analyzing the rich data she collected through her bevy of studies. I would like to thank Dr. Irwin Epstein for all his superb mentorship over these many years. Lastly, I would like to thank my colleagues at Pace University for their contributions, especially Dr. Joanne Singleton and Dr. Lillie Shortridge-Bagget, who exemplify the cutting edge mentality, intelligence, and work ethic, that move a field forward.

Contents

Preface

The primary purpose of this book is to empower the user to conduct a legitimate and effective data analysis project (a quantitative research study) from the point of conceiving of a research question, through statistical analysis, and to the completion of the project, in a professionally formatted (APA, AMA) manuscript that presents the study results. This book is also written to illustrate an effective method of interpreting published quantitative research in a manner that will allow the reader to understand, assess, and critique the statistical procedures used in a study. In short, this book is written to enable the reader, who may be a complete stranger to research and statistics, to effectively conduct, read, understand, and intelligently evaluate research studies that employ data analysis methods.

The function of this book is clear and simple. There is no pretense of literary greatness, nor does the book exhibit any unusual scholarship on my part. Rather, it simply presents data analysis in a clear and comprehensible format. Specifically, data analysis is presented as a series of seven logical, simple, and understandable steps. Many learning techniques and examples are presented that make data analysis conceptually accessible and relatable to the reader, while demonstrating that he or she does not need to be intimidated by statistical research.

Moreover, this book does not only describe the process of data analysis, but applies it. Specifically, *The 7 Steps of Data Analysis* model is applied to complete two data analysis studies for two reasons. First, these studies are presented to illustrate the many steps, decisions, and challenges encountered when conducing a data analysis study. Second, these studies act as templates for the reader to follow when implementing his or her own research activities. Primarily, the reader can plug in variables from his or her own research project and follow the process of analysis as illustrated by the studies herein. This book also applies *The 7 Steps of Data Analysis* model in assessing already published quantitative research. Essentially, every effort is made to facilitate the reader in gaining an aptitude in not only understanding these materials, but also in applying them in real world situations.

As you read this book carefully, make an effort to absorb the materials, and persistently practice the principles and formulas within, and you will be amazed at your grasp of statistical procedures and how they are applied together to conduct a data analysis study. How can I be so sure that the practice of these materials will produce such results? The answer is, for many years I have taught these techniques to graduate students in Master's and Doctoral level statistics and quantitative research courses, as well as a great many private clients, while carefully noting the operation of these methods in their learning and quantitative research skills. I found that the principles presented in this book worked efficiently over a long period of time on a consistent basis. Of course there were revisions over the years, however, the techniques outlined in this book are the finished version of the system.

In my writings, including my regular statistical newsletter *The StatsWhisperer*™, in my Webinar programs, and in lectures presented across the country, I have taught these same scientific yet simple principles of data analysis. Many have read, listened, and practiced these methods to gain proficiency in quantitative analysis. Because several of these individuals have requested that these techniques be put into book form so that the system can be better studied and practiced, I am publishing these materials in the current book titled *The 7 Steps of Data Analysis*. I would like to point out that the techniques contained in this book are not my invention. They are techniques that we, as data analysts, have for years worked hard to cultivate, test, practice and perfect. However, this is the first statistics book that presents these procedures in a practical direct-action step-by-step format as a means of facilitating effective learning and encouraging profound success.

William M. Bannon, Jr.

PART 1

INTRODUCTION

1.1 The Background of this Textbook

While this text is written as a general presentation of the methods needed to produce a quantitative study, it may be helpful to understand the initial motivation for producing the materials. Subsequently, in this section we will briefly describe the needs, ideas, and decisions that inspired this text. We will also describe the components of data analysis, as well as how the materials are organized in this text to teach you this process.

1.1.1 Who Was this Book Written for?

This book was originally written for people who now find themselves in a somewhat peculiar situation, the same one I found myself in many years ago. The situation to which I am referring is graduate school. Often a graduate student enrolls in a program to study a subject other than statistics, but finds that in order to graduate his/her chosen program he/she must learn, understand, and even conduct data analysis. This is a common occurrence in clinically-based advanced degree programs, including programs in medicine, nursing, and social work. Understanding and properly using statistics becomes especially important when the program requires the completion of a thesis or dissertation based on a data analysis project.

Regarding my own situation, I enrolled in a Masters Degree in Social Work (MSW) program with the intention of becoming a clinician. The first semester of this program included a required foundational course in Social Work Research and Statistics. Within the first month of beginning this course I noticed that I was keenly interested in the subject matter. The professor of the course told me I had been bitten by the "research bug." Soon after realizing that the research bug had bitten me, I noticed that most other students had *not* been bitten, but were generally annoyed by the research bug.

I noticed within the program at large, not every graduate student was getting a great deal of fulfillment or enjoyment from their study of research and statistics. Certainly, most were not grasping how the statistical procedures being taught could be organized toward completing a data analysis study. This left students anxious and frustrated, with a negative attitude toward research and statistics, which then prompted me to ask the question: *How can I make data analysis more understandable, meaningful, and enjoyable for advanced degree students in clinically-based programs?*

1.1.2   The Data Analysis Cycle of Inner-Knowledge

After years of learning data analysis and pondering the experience of my fellow students, I came up with an answer: **Break the Data Analysis Cycle of Inner-Knowledge**. You may wonder not only what the *Data Analysis Cycle of Inner-Knowledge* is, but also how I arrived at it. Let me explain the process. Early on, even though I graduated from a clinically focused program, I decided to become a statistician. Years later, after working on many studies, I realized something extraordinary.

It seemed that every successful statistician and data analyst had to figure out the same necessary steps and procedures needed in order to produce a professional level data analysis study. This deduction was largely achieved by becoming an efficient data analyst myself, through a combination of advice from mentors, learned coursework, and trial and error. Soon after developing this understanding, I recognized that there was a peculiar cycle where this knowledge was kept in a sort of inner circle of statisticians and data analysts. Specifically, what I recognized is that:

Once a statistician or data analyst figures out how to conduct an effective data analysis, he/she largely continues to apply this knowledge in his/her craft, but almost never thinks to share this knowledge with anyone else. Thus, I recognized what I termed *The Data Analysis Cycle of Inner-Knowledge*.

I realized that although not intentional, there was a key piece of information privy to a relatively small circle of data analysts and statisticians, but not to the mass of professionals in the research and practice world. It was as if one had figured out the secret to becoming rich, but was too busy managing and applying his/her money to remember to share the secret with people who were not yet rich.

I also realized that if this knowledge was published and related efficiently, it could help a great many professionals become effective data analysts. Essentially, the years of mentor advice, coursework, and trial and error needed to become an effective data analyst could be delivered to these professionals rather quickly and effectively. Furthermore, the mass of graduate students who view statistics as a conglomeration of confusing procedures would realize that statistical tests are a useful part of a larger process of data analysis. Additionally, those students who need to complete a data analysis project for a thesis or dissertation program requirement will have an instructional roadmap for doing so.

Thus, the goal of this book is to present, review, and share the seemingly arcane stepwise process of data analysis in a new and interesting manner that is easy to understand, and even more importantly, easy to implement. The key to this goal can be expressed in one word: *essentials*.

1.2 The Key is Knowing the Essentials

The key to conducting data analysis effectively and efficiently rests with the realization that *the mind retains a limited number of essential facts*. Perhaps it is just human nature or a quality of the mind, but the brain seems to break large concepts down only to retain a small number of essential facts, no matter what the topic. For example, a car has hundreds of working parts. Most people are unfamiliar with most of them, but they can tell you to use a car you must first insert the key into the ignition and turn it to start the engine. Most people may not be able to describe the structure of the US government, in terms of the number of senators and congressmen therein. However, most people can tell you the US government includes only one president. Even if an individual learned each moving piece of a car and a detailed description of the US government, several years later, he or she would likely have forgotten most of these specifics. However, he or she would likely still remember that you need a key to start a car and that the US government has a single president. Thus, the mind retains a few essential facts upon a topic.

If we accept the premise that the mind will most likely retain a small number of essential rules, then perhaps the most effective method of learning data analysis would be to specify and review the *essential guidelines of data analysis*. Certainly in my experience, knowing these guidelines has been the key to making data analysis understandable, implementable, and enjoyable. Thus, in this text we will identify, review, and apply *The 7 Steps to Data Analysis*, which will provide a foundation for learning, interpreting, and conducting quantitative research.

1.3   Creating a Foundation to Build Upon

Obviously, data analysis is an intricate process that takes years to master. The purpose of this book is not to turn you into a master of data analysis overnight, but to lay a foundation for gaining an ever increasing mastery of the subject. The text is predicated upon the belief that if one develops an essential foundational knowledge of data analysis, one can always add to that knowledge. However, if one does not develop an understanding of the fundamentals, then there is nothing to build upon.

Therefore, this text does not go into great detail about every small nuance regarding each topic mentioned. I have found that this approach makes a text exceedingly dense and overshadows the central message being presented. Instead, this text covers the major topics and concepts associated with data analysis. Readers are certainly encouraged to gain a greater depth of knowledge regarding these topics after completing this text. In short: *I try not to focus too much on the details, so the essentials can be clearly presented.*

One of the biggest challenges in learning data analysis is not that there is an unavailability of materials on a given topic. Today, most topics are covered rather well by online sources. The challenge is that most people are unaware of what topics they need to gather information upon. For example, in order to determine how many participants are needed for a data analysis study, you would conduct a procedure known as a power analysis. If you know this fact, you can identify many resources that will explain this procedure.

However, if you do not know the term for the procedure (i.e., power analysis), you may have difficulty gathering information on the topic. Worse yet, if you are not informed of the procedure and related concepts, you may not even realize that there is a requisite number of study participants for a data analysis study. Through presenting the major facets and facts of data analysis, this text will provide you with the essential terms, concepts, and basics you need to know to understand and/or conduct a data analysis study. Furthermore, this book was printed with wide margins to give the reader the opportunity to enter by hand any important information pertinent to their learning and research activities that may not be mentioned in the text.

1.4   Layout of the Text

You might notice that the layout of this book is dramatically different from any other statistical textbook, which is done intentionally. The materials are laid out in such a way that they can be applied sequentially as you conduct a data analysis study. Along with the conceptual information, the text includes instruction on how to conduct each procedure in statistical software, as well as how to effectively interpret the statistical output. The statistical software program SPSS is used, as that seems to be the most popular and widely used data analysis program.

I have attended many lectures where I felt the information presented was spectacular and I was excited to apply the information once I got back to my desk. However, I often found that once I sat down to apply the newly learned information, I could not figure out how to do so. Subsequently in this text, information is not only presented, but laid out in a manner for direct application.

1.5   What is Data Analysis?

This book has been written with the intention of instructing the reader on how to learn, understand, interpret, and conduct data analysis. That may prompt you to ask, what is data analysis? If you ask the average person what data analysis is, you are likely to get the response "statistics." While statistics are the primary tool used in data analysis, statistical tests and procedures actually only comprise a segment of data analysis. Essentially, data analysis is applying a series of statistical tests and procedures in a specific stepwise progression in order to examine a dataset. In other words, data analysis is taking a set of tools (statistical tests/procedures) that when applied in a certain order (*The 7 Steps of Data Analysis*) reveal the message(s), lesson(s), and answer(s) the data have to tell us. However, the foundational nature of this textbook suggests data analysis might be foreign to some readers. Therefore, perhaps a more effective discussion might focus on how the components of data analysis may be similar to another process already familiar to the reader, such as making a cake.

1.6   The Components of Data Analysis

In short, the components involved in making a cake are almost identical to those in conducting a data analysis study. Therefore, to make the lesson easier, we will present how the components of making a cake are related to the components of conducting a data analysis study. My hope is that this parallel will make the concept of data analysis much more "digestible." Subsequently, let's first talk about the components of making a cake. Just to keep it simple, let's say you need three things to make a cake. Specifically, you need the:

1) **Cake Recipe**: An *outline of the steps to follow* toward making that cake, which is simply a "to do" list presenting a series of steps that must be taken in a precise order.

2) Cake Ingredients: The *materials* necessary to produce the cake.

3) Cooking Utensils: The *tools* to transform the ingredients into the product (cake).

To conduct a data analysis, you need the equivalent three things, which are the:

1) 7 Steps of Data Analysis: An *outline of the steps to follow* toward conducting a data analysis study. Like a recipe, I suggest *The 7 Steps of Data Analysis* is also simply a "to do" list presenting a series of steps in a precise order (a **recipe** for a study).

2) Study Data: The *materials* necessary to produce a data analysis study (**ingredients** for a study).

3) Statistical Tests and Procedures: The *tools* to transform the data into the product, which is the completed data analysis study (**utensils** for a study).

We will clarify these relationships and parallels in the next part of this section.

1.6.1   The Cake Recipe & The 7 Steps Of Data Analysis

As we suggested, you could think of a completed data analysis study like a completed baked good, like a cake. The data analysis study and cake are both finished products that were created by following a "to do" list. For example, in the box labeled **To Do List: Making a Cake** within **Figure 1.1**, the **Cake Recipe** is presented, which is a list of things that must be done to produce the product (a cake). In the box labeled **To Do List: Data Analysis Study**, **The 7 Steps of Data Analysis** is presented, which is a list of things that must be done to produce the final product (a quantitative study). Within **Figure 1.1**, you will see a side by side comparison of how the **Cake Recipe** and **The 7 Steps of Data Analysis** are each a list of steps that must be followed and completed in a certain order to produce the respective final product.

Figure 1.1   The *to do* lists: The cake recipe and The 7 Steps of Data Analysis model

TO DO LIST: MAKING A CAKE	TO DO LIST: DATA ANALYSIS STUDY
Cake Recipe	**The 7 Steps of Data Analysis**
1) Preheat oven to 350 degrees	1) Create a Study Map
2) Grease and flour a 9x9 inch pan	2) Data Entry
3) Cream together sugar and butter	3) Check Data Integrity
4) Beat in the eggs	4) Univariate Analysis
5) Stir in the vanilla	5) Bivariate Analysis
6) Add flour and baking powder	6) Multivariate Analysis
7) Stir in the milk until smooth	7) Write-up & Report
8) Bake for 30 to 40 minutes	

Of course each step within **The 7 Steps of Data Analysis** model involves many other steps and considerations. In other words, each of the seven steps is a representation of many other smaller and more detailed steps. However, every data analysis study can be approached by categorizing these smaller and more detailed steps into this *to do* style list toward completing an efficient statistical analysis.

1.6.2   The Cake Ingredients & The Study Data

The cake is created using materials, known as ingredients. The word "ingredients" is really just a catch-all phrase. The materials known as ingredients have specific names and qualities, such as butter and flour. Furthermore, different cakes require different types, amounts, and combinations of these materials. In other words, each cake has specific needs regarding the necessary ingredients used to make that cake. For example, within **Figure 1.2** the box labeled **Materials: Making a cake**, describes the necessary ingredients for making this specific cake, such as butter, eggs, sugar, etc. However, a different cake recipe might call for replacing these ingredients with alternatives.

The data analysis study is conducted using materials, known as data. The word "data" is really just a catch-all phrase. The materials known as data are structured as variables (for a fuller discussion of variables see section **3.4 Study Variable Type**). Different quantitative studies require different types, amounts, and combinations of these variables. For example, as listed in the box labeled **Materials: Data Analysis Study** within **Figure 1.2**, a study might include one dependent variable (*Happiness*), one independent variable (*Do You Live with a Dog or a Cat?*), and two covariate variables (*Education Level* and *Income Level*).

Figure 1.2   The needed project materials: Cake ingredients and study data (variables)

MATERIALS: MAKING A CAKE	MATERIALS: DATA ANALYSIS STUDY
The Cake Ingredients	**The Study Data (Variables)**
1) 1 Cup of White Sugar	1) Dependent Variable: Happiness
2) ½ Cup of Butter	2) Independent Variable: Do You Live with a Dog or a Cat?
3) 2 Eggs	3) Covariate Variable: Education Level
4) 2 Teaspoons Vanilla Extract	4) Covariate Variable: Income Level
5) 1 ½ Cups All Purpose Flour	
6) 1 ¾ Teaspoons Baking Powder	
7) ½ Cup of Milk	

However, like cake ingredients, these factors will vary from project to project. For example, another study may require more or less independent or covariate variables. Additionally, a study may require different types of variables. For example, the variable *Happiness* might be replaced with the variable *Depression* as the dependent variable incorporated in the study.

1.6.3 The Cooking Utensils & Statistical Tests

After the cake recipe and ingredients have been selected, specific tools known as cooking utensils are used to relate the ingredients to one another. Relating the ingredients successfully will result in a sound final product (a cake). The rules for relating the ingredients are clear. For example, within the box labeled **Tools Relating: Ingredients** in **Figure 1.3**, the first line instructs to relate (i.e., *cream*) the ingredients *sugar and butter* you would use a cooking utensil known as a *Hand Mixer*. The next line informs to relate (i.e., *beat*) the ingredient *Eggs Into The Overall Mixture*, you would use a cooking utensil know as a *Whisk*. Of course, there are many other types of cooking utensils not mentioned, but you get the idea.

Simply put, just as there are cooking utensils to relate the cake ingredients, there are statistical tests to relate study variables to one another. For example, the first line within the box labeled **Tools Relating: Study Variables** in **Figure 1.3**, describes to relate the variables *Happiness* and *Education Level*, you would use a statistical test known as an *ANOVA*. The second line describes to relate the study variables *Happiness* and *Do You Live With a Dog or a Cat?* you would use a statistical test known as a *T-test*. Relating the variables with the correct statistical test will result in a sound final product (a quantitative study).

Figure 1.3 The needed project tools: Cooking utensils and statistical tests

TOOLS RELATING: INGREDIENTS		TOOLS RELATING: STUDY VARIABLES	
Ingredients Related	**Cooking Utensil Used**	**Study Variables Related**	**Statistical Test Used**
Cream Together Sugar And Butter	Hand Mixer	Happiness & Education Level	ANOVA
Beat The Eggs Into The Mixture	Whisk	Happiness & Do You Live With a Dog or a Cat?	T-test
Stir Vanilla In With Other Ingredients	Large Spoon	Happiness & Income Level	Correlation

1.7 Why Statistics Are Awesome

In the last section, we described how statistical tests and procedures are the primary tools used in data analysis. But, just what are statistics and why are they awesome, as the title of this section suggests? The response to both questions is: statistics are tools that help us **understand the reality around us in a way we could not otherwise!** Statistics offer us a means of unearthing knowledge about the world we live in that otherwise would remain hidden. Specifically, statistical procedures are methods of measuring the world around us as variables (e.g., *Happiness* or *Education Level*) and then testing the association(s) between those measured variables (e.g., *Is Happiness Associated With Your Highest Level Of Education?*). These measured variables are largely undetectable by the five senses of touch, taste, smell, sight, or hearing. However, we can measure these variables empirically by producing a portrait of these unseen aspects of the world around us via statistical methods.

For example, let's say you are among a crowd of 100 people at a restaurant. While you are waiting for a table, you begin to wonder if among this group, the people *With Graduate School Degrees* have a higher level of *Happiness* relative to all other patrons (those who *Do Not Have A Graduate School Degree*). Could you use your sense of touch, taste, smell, sight, or hearing to effectively observe if the relationship between these variables exists? Not really, as you cannot touch, see, hear, taste, or smell one's *Level of Education* or *Happiness*. However, if you provided each table with an empirical measure of *Happiness* and level of *Education* you certainly could identify these factors! In this way statistics are like a sixth sense!

1.7.1 Applying Statistics to Home and Work Life

Statistics can also be used to understand the reality of your home life, as well as your work life. Taking your home life as an example first, let's say you are married and you notice that each time you compliment your spouse in the morning he/she cleans the house that evening. Soon you get tired of wondering if there is an association between this complimenting and cleaning scenario, so you begin to gather data toward producing a better picture of what is really going on. Over the next 100 days you compliment your spouse once in the morning for 50 days and do not compliment him/her on the other 50 days. You then count the <u>number of times your spouse has cleaned the house</u> on the days that you <u>did</u> and <u>did not compliment</u> him/ her.

You table these data in a manner similar to **Table 1.1**, which describes on the mornings you gave a compliment to your spouse he/she cleaned the house 80% of the time. On the mornings you did not give a compliment, your spouse cleaned the house 5% of the time. Now you have actual values, or statistics, that better reflect the reality of your home life. The data reflect that your spouse seemed to clean the house on a much higher percentage of the days you paid them a compliment (80%) relative to the days you did not (5%), which suggest a significant relationship between the two occurrences.

Table 1.1   The relationship between *gave a compliment* and *cleaned the house*

		Cleaned the House		
		Yes	No	Total
Gave a	Yes	80%	20%	100%
Compliment	No	5%	95%	100%

An example like this could just as easily be generalized into one's professional role. For example, if you were a nurse and had a patient that did not like to take his/her daily medication, you could set up the same experiment to determine if your complimenting the patient had an association with the percentage of time he/she took his/her medications (please see **Table 1.2** below).

Table 1.2   The relationship between *gave a compliment* and *patient took medication*

		Patient Took Medication		
		Yes	No	Total
Gave a	Yes	80%	20%	100%
Compliment	No	5%	95%	100%

Again, these same numbers indicate the patient took his or her medication on a much higher percentage of the days you paid them a compliment (80%) relative to the days you did not pay them a compliment (5%). This would again suggest a relationship between complimenting the patient and his or her taking medication. So you can see that without exaggeration, statistics allow us to see reality in a deeper and more meaningful way than we could otherwise.

Many of us have heard and agree with the statement by Socrates "The unexamined life is not worth living." However, while the statement is widely circulated, the knowledge of *how* to examine one's life is not, which may be a cause of concern not only in your personal life, but your professional life too. For example, suppose you are a physician who believes it is not worth treating patients unless you can examine how well your treatments work. In that case, determining *how* to examine this facet of your professional life would be a critical challenge. Statistics and data analysis are one of the few methods available that can provide this *how*.

1.7.2 Curing the *Who's The Boss* Syndrome

Meaningful research questions often surround us, but frequently these important potential areas of study are left unaddressed. To remind us that these topics of study exist, we might benefit from a catchy phrase that may potentially stick in our minds to help us keep on the lookout for these unaddressed research questions. Therefore, I have ascribed these potential areas of study to be part of the *Who's The Boss* Syndrome.

In the 1980s, the sitcom *Who's the Boss* (1984-1992) debuted. The program aired almost every Thursday night for eight years. Each time the program aired, the opening credits would present the question *Who's the Boss?* as the title of the show was revealed. This question was presented weekly for the eight year duration of the show (many more times if you count syndication). However, when the show ended in 1992, the viewer was never told of the cast characters, just who the boss was. Thus, the same question was posed for nearly a decade, but never answered. In fact, many viewers seemed to even forget a question was being asked and did not expect an answer. Although this development may not seem very consequential, this very thing often occurs in many professional areas of study and must be diligently guarded against.

The *Who's the Boss Syndrome* occurs when we are faced with important questions regularly (e.g., every Thursday night on the ABC television network) that seem to perpetually go unaddressed and unanswered. Some of these questions might be light, but some of these questions are profound and essential. For example, a certain medical treatment might be widely used within a certain population for decades, but the effectiveness of this treatment might never have been examined. Thus, there may be a perpetual question of *Is this treatment effective?* that might be pondered year after year without ever being answered. This would be the more serious side of the *Who's the Boss Syndrome*.

In light of such situations, we might suggest that rolling questions are generally not acceptable. Specifically, if an important question exists, we must find an effective means of answering it. In other words, perhaps it is not permissible to continuously ask the same important question regularly and not produce a meaningful answer. I found that statistics and data analysis are a superior means of producing meaningful answers to such questions. Subsequently, a viable solution to the *Who's The Boss* Syndrome is applying statistics and data analysis toward answering the important questions with which we are faced.

1.8   Applying the Materials

Learning new and interesting information can be a worthwhile pleasure. However, most people have had the experience of learning seemingly useful information, only to discover that he or she can't figure out how to apply it! It seems that at some point we all realize that there is often a huge gap between *knowing great information* and *knowing how to apply that great information*. Therefore, in this textbook, we will move beyond the presentation of knowledge and into the application of knowledge! Specifically, we will apply *The 7 Steps of Data Analysis* in three instances.

First, in **Part 4**, we will conduct sample study one, which is a quantitative research study examining a continuous dependent variable. Specifically, sample study one examines if the continuous dependent variable *Happiness* is related to the independent variable *Do You Live with a Dog or a Cat?*, as well as the covariate variables *Education Level* and *Income Level* at a statistically significant level. Next, in **Part 5**, we will conduct sample study two, which is a quantitative research study examining a dichotomous (a categorical variable with two response categories) dependent variable. Specifically, sample study two examines if the dichotomous dependent variable *Happy* (Yes or No) is related to the independent variable *Do You Live with a Dog or a Cat?*, as well as the covariate variables *Education Level* and *Income Level* at a statistically significant level. Lastly, in **Part 6**, *The 7 Steps of Data Analysis* model is applied to assessing the quality of a published quantitative research study.

Please note that **Parts 4**, **5**, and **6** are meant to serve as templates, which can be generalized to data analysis studies of one's own. For example, if an individual had a data analysis study to complete that incorporated a continuous dependent variable, he or she could plug in that variable within **Part 4** (our data analysis study examining a continuous dependent variable) and follow all the steps thereafter toward completing his or her own study.

However, before applying *The 7 Steps of Data Analysis*, it is essential to discuss how these steps compose a needed and timely model of data analysis. Therefore, in **Part 2** we will discuss the concept and utility of a model of data analysis, as well as the qualities such a model should reflect. Additionally, in **Part 3**, we will share some fundamental concepts that the reader should be aware of prior to approaching data analysis.

PART 2

A MODEL FOR CONDUCTING STATISTICAL RESEARCH

2.1 What Will This Section Tell Us?

This section will discuss not only how *The 7 Steps of Data Analysis* can serve as a model to follow toward effectively conducting statistical research, but also how a model of data analysis is essential to the field of statistical research at this juncture of professional evolution. Specifically, today the field of statistical research is experiencing an unprecedented demand where many professionals across several disciplines desire to learn, understand, and apply the technology involved with quantitative research. In the text below, we will discuss how *The 7 Steps of Data Analysis* can be applied as a tool to share the technology of statistical research with the wide audience interested in the materials.

2.2 The Evolution of Professional Models: Taking it to the Streets

The natural evolution of a profession includes the development of dependable models that instruct on how to use the technology to produce outcomes in that field. For example, we mentioned earlier that in the culinary arts, a baker follows a model, known as a *recipe*, to produce a cake. Certainly, as baking became a profession, the professionals within that field began to develop and refine these models (recipes), where the professional went from having no formal models to a great many.

A second evolutionary point within a profession is marked by the period where the technology within that profession, formerly only accessible to members of that field, becomes accessible, understandable, and implementable to people outside that field through a crossover tool. In other words, there is an evolutionary point within each field where through an effective model, the processes and outcomes produced within that profession are made available to people outside that profession. One could say that each profession experiences an evolutionary point where the technology is taken from the inner circles of that profession to the streets via an effective model.

Consider the field of computers as an example. The first computer is estimated to have been invented in the early to mid-1800s. For approximately 150 years, the general public coexisted with, but was largely independent of computers. During this period the comprehension and application of computer software and hardware rested almost exclusively within the professionals operating in that technological field. However, in the late 1970s, the field reached an evolutionary point where computer technology would now be made available to individuals outside the field of computers. Specifically, Steve Jobs, along with Apple co-founders Ronald Wayne and Steve Wozniak, began to develop the first personal computer (PC), such as the Apple 1 and so on, which would allow members of the general public to understand, implement, and even develop, computer technology.

As a second example, consider the professionals that prepare French cuisine. The foundations of modern French Cuisine, referred to as Haute cuisine or High cuisine, began in the 1760s with Chef La Varenne. For approximately 200 years, America coexisted with Haute cuisine, yet the only access Americans had to this discipline was eating at restaurants. In other words, the models used to produce the dishes in this field were largely not accessible to the American public. In the 1960s, Haute cuisine reached an evolutionary point were the technology in that field would now be made available to people not operating directly in that specialty area. Specifically, in the 1960s, Julia Child successfully published a series of models (i.e., recipes in her book *Mastering the Art of French Cooking*) that made the preparation of Haute cuisine accessible to all.

As a final example, consider American football. Professional American football had been around since the mid-1800s. However, from that time to the first half of the 20th Century, American football seemed somewhat foreign to the American public. Then in the 1960s, the game became understandable, replicable, and embraced by the Americans. This point of evolution for the game was ushered in by Vince Lombardi of the Green Bay Packers. Among his many contributions, the coach was able to clearly communicate the fundamentals of playing the game. Specifically, the coach illustrated his football plays (his models for scoring points) so clearly that each play was understandable and implementable by everyone from the top player in pro ball, to high school and college teams, to young children playing in a sandlot.

Across all three examples, you might notice a common trend where the quality of the lives of the people outside of those professions was dramatically enhanced because the technology within these fields had been made available to them. For example, consider how the lives of the earthly population have been enhanced by the PC. For approximately 150 years the possibility of a computer accessible to the general public existed. However, it was not until Steve Jobs and his Apple cofounders developed a crossover tool that benefited the lives of people outside the field of computers by making this technology available to and usable by a wide audience. Today, the access to computer technology facilitated by the PC enhances the lives of a great many people in a great many ways.

2.3   The Evolution of Statistical Research: Taking It to the Streets

Modern statistics is still a somewhat young field. The foundation of modern statistical methods is largely seen as developing in the mid-1700s. Thus, modern statistics began to flourish somewhere in between the period modern French cuisine appeared and the time computers and American football came about. I mention these fields in tandem, as it seems similar to our first three examples, that the field of statistics has reached that second evolutionary point where the specialized technology within the field is now becoming accessible to people outside the field.

However, the arrival of this evolutionary point within the field of statistics developed in a different manner relative to the other three examples. Specifically, in the first three examples, a crossover tool(s) was first developed (the PC, the cookbook, the coach's plays), then the technology was made available to a wider population. Prior to the development of these crossover tools, there was not a pressing need to develop these models so the technology within each field could be shared. For example, there was not a pressing need for a wider population to have a PC, learn French cooking, or play American football.

Regarding statistical research, the need for a crossover tool to make the technology accessible to a wider population has somewhat preceded the availability of the tool. Seemingly, today there are more professionals, who are not statisticians or data analysts, across more fields attempting to learn, teach, conduct, and interpret statistical procedures and research than any other time in recorded history.

A short time ago, the analysis and interpretation of statistical procedures were left largely up to the respective statistician in that field. For example, in a hospital or school of medicine, the person employed as the biostatistician might have had sole responsibility in conducting statistical analysis and interpreting the findings in layperson's terms to the medical staff and faculty. In other words, at that time it was acceptable and even expected that the biostatistician was the only staff member who could read and speak the language of statistics.

These days the biostatistician has a lot of company. Specifically, many professional fields are evolving to a point where a significant percentage of staff is expected to be able to read and speak the language of statistics. For example, in the same hospital or school of medicine mentioned above, now along with the biostatistician, the nurses, social workers, doctors, and many other staff and faculty are expected to be able to understand, evaluate, and conduct statistical research. Of course, along with the expectation to learn statistical research, a great deal of resources have been developed in the form of academic and other training programs as a means of molding professionals into outstanding researchers.

This might lead us to ask, why are the heads of so many professional fields asking members to become proficient in statistical research? Essentially, it would seem that the leaders within these professions recognize that this crossover technology, just like the PC, has a great potential to enhance the quality of life for not only the professionals within these fields, but all persons served by these professions. Put plainly, today a wide audience of professionals (and beyond) regard statistical research as valuable to the point where they believe members should be able to understand, interpret, evaluate, and even conduct statistical research. This is a remarkable development, which is gaining momentum.

A crucial resource to support this movement would be a key model that would make the technology of statistical research accessible to a wider audience outside the field of statistics. The crossover tool to bring the technology of statistical research to a broader audience outside the field of statistics already exists, but is seldom articulated. Specifically, after the necessary level of professional education and experience, every statistician and data analyst realizes there are essentially seven steps involved in a data analysis study. I suspect a great deal of the time, this realization is largely unconscious, which is why the information is seldom articulated. Nonetheless, the fundamentals of statistical research can be addressed through approaching data analysis as a seven step process. This process is expressed as *The 7 Steps of Data Analysis* model presented in this text.

In this section we will present <u>how</u> *The 7 Steps of Data Analysis* model acts as a crossover tool that will make learning, teaching, interpreting, evaluating, and applying statistical procedures accessible to a wider audience. Specifically, in this section, we will describe how elements of this model break down the data analysis into manageable steps, making the process clear and implementable by persons who are not necessarily statisticians and data analysts.

2.4   What Elements Should a Model of Data Analysis Reflect?

Across professional fields, effective models have common elements. Therefore, when assessing *The 7 Steps of Data Analysis*, a solid method of quality assessment would be to compare this model with the characteristics of a near flawless model employed in another professional field. For example, we might examine the elements of Coach Lombardi's main offensive football play (his main model for scoring points), known as the *Power Sweep*, and discuss how the characteristics of this model would benefit a model of data analysis. Before we discuss the elements of Coach Lombardi's main football play, we will explain the maneuver, as well as the essentials of American football.

The Fundamentals of American Football. **Figure 2.1** presents that in an American football game there are 22 people playing on the field. On one side (the lower portion) there are 11 offensive players trying to get the football into the end zone to score a touchdown (points). On the other side (the upper portion) there are 11 defensive players trying to stop the offensive team from scoring that touchdown. The offensive team has plays that involve passing and/or running the football, as a means of getting past the defensive team. While passing and running the football are clear objectives of the game, the biggest fundamentals are blocking and tackling. Blocking generally refers to members of the offensive team blocking members of the defensive team, so other members of the offense can run and pass the ball. Tackling refers to members of the defensive team fighting off those offensive blocks to tackle whichever member of the offensive team is running or passing the football.

The Power Sweep Football Play. **Figures 2.1., 2.2** and **2.3**, illustrate the utter simplicity of the power sweep play. In **Figure 2.1**, you will see the formation (how and where each player is set up) of the teams right before the power sweep is run. **Figure 2.2** presents the power sweep play being run. Essentially, circle #8, the quarterback, hands the ball to circle #11, the halfback, who begins to run to the left. At the same time, circle #4 and circle #6 (the 2 offensive guards) leave their spots to run in front of the halfback (circle #11) to block any defensive players (the squares).

Figure 2.1   The football formation before the power sweep play is run

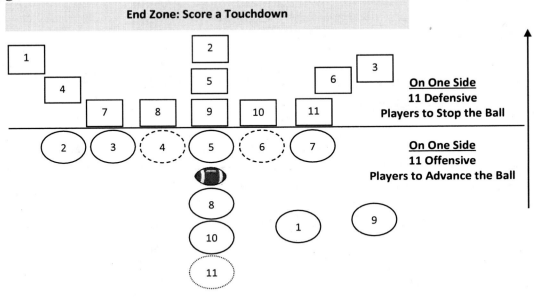

Figure 2.2   The power sweep play being run

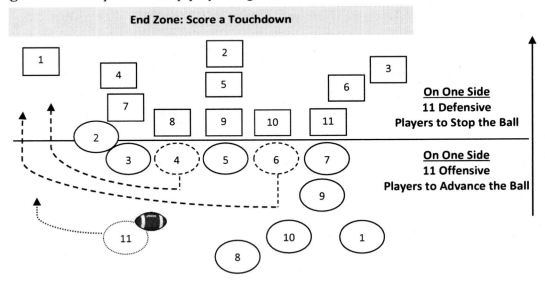

The final image, **Figure 2.3** illustrates how the blocking and running result in the offensive ball carrier successfully moving down the field to score a touchdown. These are the essentials of the power sweep play.

Figure 2.3 The result of power sweep play

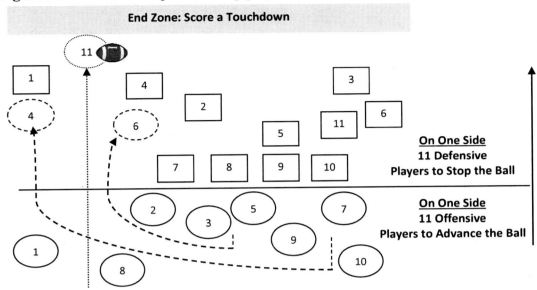

In the next section, we will discuss how a few simple characteristics reflected within this play, are the key traits that would make a model of data analysis successful.

2.5 Shared Elements of Successful Models

In this section we will review the essential shared elements between the most successful professional model (play) in football, Coach Lombardi's *Power Sweep*, and *The 7 Steps of Data Analysis* model. Through relating the two, we will illustrate how *The 7 Steps of Data Analysis* model possesses sound features that will facilitate successful use and application.

2.5.1 Fundamentally Sound

At the most basic level, both models are fundamentally sound. A model that does not consider the necessary essentials cannot produce consistent results. For example, as mentioned earlier, the main objectives of an offensive play in American football are blocking, as well as running or passing the football. These fundamentals are all reflected in the power sweep play by necessity. For example, if the play addressed the importance of running the football, but ignored a blocking strategy of the defensive players, the model would not be successful no matter how well the ball was run.

Like the *Power Sweep* football play, the *7 Steps of Data Analysis* model is also fundamentally sound. Specifically, this model makes a clear consideration of all the rudiments of statistical research. For example, the steps refer to a planning stage (Step 1), preparing the data properly (Step 2), assuring data are appropriate for analysis (Step 3), analyzing the data (Steps 4-6), and reporting the findings in the proper style (Step 7). Failing to consider any of these steps would be neglecting the fundamentals of data analysis. For example, if the appropriateness of the data was not examined in Step 3 the entire analysis could very well be unsound and misleading.

2.5.2   Conceptually Clear and Understandable

If a model was a superhero, conceptual clarity would be its superpower. The level of conceptual clarity within a model will determine how effectively the professional technology will be transmitted and understood by those using that model. For example, we saw that the power sweep could be effectively expressed in three simple images, even to people unfamiliar with American football. This is not the case with most plays in American football. Many plays would be rather complicated even to people acquainted with the game. However, the simplicity of the power sweep represents one remarkably important possibility. Specifically, after viewing these three images, even people never exposed to American football could likely understand and perform the play. This prospect of immediate application of knowledge represents the best chance one will eventually master the materials being learned.

To effectively learn any new information, one needs more than simple instruction. To ground his or her learning, one needs to apply the information learned as soon as possible (in a contextually appropriate manner of course). Often, to the degree that information is learned, but not applied, a learner will be more likely to forget and never make use of his or her learning. Therefore, a clear model that facilitates an immediate application of information represents the most promising chance that the materials will be grasped, applied, and eventually mastered. Likewise, the more complex the model, the more likely the information will not be understood nor ever applied.

Essentially, the clear and simple model becomes what I call a *Positive Gateway Activity* to the professional technology reflected in that model. For years I enjoyed working in prevention research, and became familiar with terms such as gateway drugs and gateway sexual activity. However, I also realized there are gateways into positive activities, which are often professional models. For example, suppose a young boy saw Coach Lombardi's power sweep play run in a professional football game. After being exposed to this model, the boy then runs that play himself with his friends in a park, then in high school, then college, then in professional American football himself. Note that the exposure to that model (play) was the boy's gateway to professional excellence.

Also, please note, that the play was the boy's gateway to excellence in a large part because the conceptual clarity of the model enabled the boy to look at that play, understand that play, and immediately begin to apply that play. Then through study and repetition, the boy would be able to bring his application of that model from the beginner level to the professional level. If the play was initially complicated and inaccessible, it may not have acted as a gateway to professional learning and excellence. The exact same can be said for a model of data analysis. If the model is conceptually clear, the information could act as a gateway activity that could bring model practitioners from the beginner level to the professional level of data analysis.

Optimally, the model of data analysis would be as accessible as the power sweep play, where one could see it, understand it, and immediately begin to apply it. In my work in prevention research, I have come to believe that the development of positive gateway activities is as important to supporting positive outcomes as the prevention of negative gateway activities is to dissuading negative outcomes. Thus, a model of data analysis serving as a positive gateway activity to proficiency in statistical research might be quite useful.

An Extremely Important Consideration Regarding Conceptual Clarity. In the 1980s there was a camera commercial that stated "So simple anyone can use it". Our model of data analysis should fit this uniquely profound criterion. Specifically, it is not just academics and professionals within certain fields who desire to understand statistical research, but many people beyond. For example, the family members of patients receiving medical care and the patients themselves often desire to be able to learn, understand, evaluate, and even conduct statistical research. For these *very important consumers and perhaps producers* of statistical research, language specific to certain professions might not facilitate model usage. Therefore, ideally a data analysis model should be conceptually clear irrespective of any boundaries related to an academic or professional field or institution.

Data Analysis and Dating Analysis. Because conceptual clarity is essential for a model, *The 7 Steps of Data Analysis* model is laid out in a manner that reflects much of the daily life that we or the people around us are likely experiencing. For example, you could quite easily relate the steps presented in the model of data analysis to the steps involved in a practice as ubiquitous as online dating. Please refer to **Figure 2.4**, which presents the parallels between the process of conducting a statistical research project (a.k.a., data analysis) and the process of following an online dating protocol toward identifying a person with whom you would like to go out on a date (a.k.a., dating analysis). In this section, we will now compare the parallel between data and dating analysis. You will see each process involves a series of seven similar logical steps.

In dating analysis (online dating), the first step is a planning stage where you identify the goal (e.g., find a date) of this process (Step 1), followed by entering your personal information into the online dating company database (Step 2), followed by checks to verify that you, as well as the company and other daters, have integrity (Step 3). Next, you would create your user profile and examine other user profiles (Step 4), toward identifying significant one on one relationships with other daters (Step 5). Then of all significant relationships, you would then identify the strongest relationship for a date (Step 6). Lastly, you might report the result of this process to family and friends (Step 7).

Similarly, in data analysis, the first steps are a planning stage (Step 1), data entry, into software database (Step 2), and performing checks of data integrity (Step 3). Next, a profile of each variable is created via univariate analysis (Step 4), followed by examining significant one on one relationships between variables via bivariate analysis (Step 5), and identifying the strongest variable relationship through multivariate analysis (Step 6). Lastly, the study results are reported in a proper format (Step 7).

Figure 2.4   The 7 Steps of Data Analysis and Dating Analysis Models

	7 Steps of Data Analysis	**7 Steps of Dating Analysis**
Step 1	**Study Map**	**Dating Map**
Step 2	**Data Entry**	**Data Entry**
Step 3	**Check Data Integrity**	**Check Dater Integrity**
Step 4	**Univariate Analysis** (Profiles of Individual Variables)	**Dater Profiles**
Step 5	**Bivariate Analysis** (Significant 1 on 1 Relationships)	**Identify Significant 1 on 1 Relationships**
Step 6	**Multivariate Analysis** (Strongest Relationship)	**Identify the Strongest Relationship**
Step 7	**Write-Up & Report**	**Write-Up & Report**

We will now expand on the similarity between each corresponding step within *Data Analysis* and *Dating Analysis*.

Step 1 (Data Analysis): Study Map. In Data Analysis, the first step involves drawing out a Study Map describing the goal(s) and methods of the study. The goal(s) might be to answer a certain research question(s) and/or hypothesis(es), which will require specifying certain factors including the covariate, independent, and dependent variables. For example, as depicted in **Figure 2.5**, perhaps we might want to examine which of three predictor variables (*Education Level*, *Income Level*, and *Do You Live with a Dog or a Cat?*) relate to the dependent variable *Happiness*.

Figure 2.5   The relationships examined via data analysis

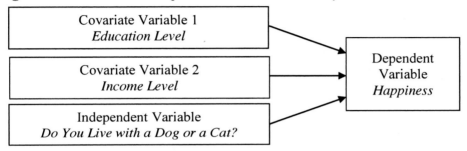

Step 1 (Dating Analysis): List of Dating Goal(s). Regarding the process of online dating you would also begin with goals and methods in mind (dating self-help books do recommend that you write these down, however, I suspect that most people make mental notes). For example, you might have the goal of finding several bachelors (or bachelorettes, but for simplicity sake, we will use the term bachelors) with whom you have a potentially significant relationship, and then select the one with whom you have the strongest relationship as a companion for a date. As presented in **Figure 2.6**, you might want to examine how three bachelors relate to you as a potential companion.

Figure 2.6   The relationships examined via dating analysis

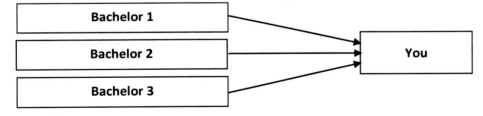

Step 2 (Data Analysis): Data Entry. In data analysis, the second step is entering all study data into a software database (data entry), and making sure all the data have been entered correctly (cleaning). For example, as presented in **Image 2.1**, you might enter the data into a software database using SPSS.

Image 2.1   The SPSS software program and data entry

Step 2 (Dating Analysis): Data Entry. In dating analysis, the second step is going online and entering your personal data into the online dating company's software database (data entry). You would also want to make sure these data are entered correctly (a.k.a., data cleaning) or else you might end up with different types of suitors than you had in mind. For example, as presented in **Image 2.2**, you might enter your information into the database of an online computer dating company such as d Harmony.

Image 2.2   The d Harmony dating program and data entry

Step 3 (Data Analysis): Checks of Data Integrity. In data analysis, the next step involves making checks of data integrity to be certain that the data are appropriate for analysis. For example, parametric testing assumes the distribution of scores (e.g., *Happiness* scores ranging from 1-5), is normal (distribution A in **Figure 2.7**). However, a distribution such as B or C would reflect a non-normal distribution and would therefore be a threat to data integrity.

Figure 2.7   Threats to data integrity: The non-normal distribution of scores (B and C)

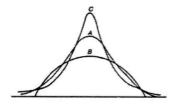

Step 3 (Dating Analysis): Checks of Dater Integrity. Similarly, online dating companies should also assess the integrity of the daters in the system to be sure each is appropriate for dating. For example, you might assume that each bachelor you have

contact with has integrity where he is at least being honest with you when responding to questions. However, if this assumption is not met, you might decide that lack of honesty (see in **Figure 2.8**) is a threat to dater integrity that makes a bachelor inappropriate for dating.

Figure 2.8 Threat to Dating Integrity: A Lack of Honesty

Step 4 (Data Analysis): Univariate Analysis. In data analysis, the fourth step involves presenting the characteristics of each individual study variable (similar to a profile of the individual variable) via univariate analysis. For example, in **Figure 2.9**, the dependent variable *Happiness* is presented. We see that of the 1-5 potential range of scores, the average (mean) score is 2.73, which is just below the midpoint (of 3.00) of the potential range of scores. Furthermore, the lowest score (the minimum = MIN) is 1.20 and the highest score (the maximum = MAX) is 4.40, indicating there are no scores that reflect the lowest or highest possible scale values of 1.00 or 5.00.

Figure 2.9   Profile of the dependent variable *Happiness*

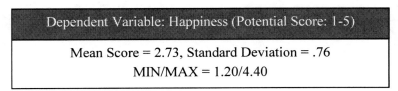

Dependent Variable: Happiness (Potential Score: 1-5)
Mean Score = 2.73, Standard Deviation = .76 MIN/MAX = 1.20/4.40

Step 4 (Dating Analysis): Dater Profiles. In Dating Analysis, the fourth step involves creating a dater self-profile, as well as examining other dater profiles. This profile lists individual characteristics, such as in **Figure 2.10**, where we see bachelor #1, Bob, is a 47 year old male who likes Italian food.

Figure 2.10   Profile of the dater *Bob*

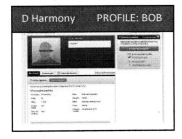

Bachelor Profile #1: **Bob**

Age: 47

Gender: Male

Likes: Italian Food

Step 5 (Data Analysis): Bivariate Analysis (Identify Significant 1 on 1 Relationships).
In Data Analysis the next step involves identifying which predictor variables have a
significant one on one relationship with the dependent variable via bivariate analysis. If
a predictor variable has a significant one on one relationship with the dependent
variable, then the predictor variable will be advanced to the next stage of analysis (**Step
6: Multivariate Analysis**). The standard indicator of a significant relationship is a
probability level of .05 or below, which is indicated by the notation $p<.05$. In **Figure
2.11**, the arrows connecting the variables *Do You Live with a Dog or a Cat?* and
Happiness indicates a significant relationship ($p<.05$). Also, the arrow connecting
Income Level and *Happiness* indicates a significant relationship ($p<.05$) between
variables. However, the arrow connecting *Education Level* and *Happiness* is $p>.05$,
which indicates a non-significant relationship ($p>.05$) between variables. Therefore, the
predictor variable *Education Level* will not be included in the next step (**Step 6:
Multivariate Analysis**).

Figure 2.11   Significant ($p<.05$) one on one relationships through bivariate analysis

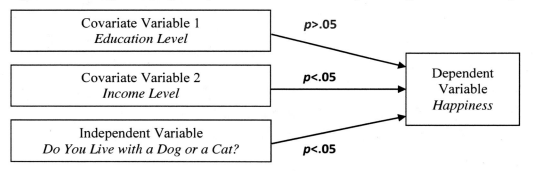

Step 5 (Dating Analysis): Identify Significant 1 on 1 Relationships. At this point, you
would have a pool of people who responded to your profile. Your next step would
likely involve identifying with whom in this pool you feel you have a significant one on
one relationship. You might then exclude the remaining applicants from further
consideration. For example, as presented in **Figure 2.12,** you may have decided you
had a significant relationship with bachelors 1 and 2, but not 3.

Figure 2.12   Identifying significant ($p<.05$) one on one relationships with daters

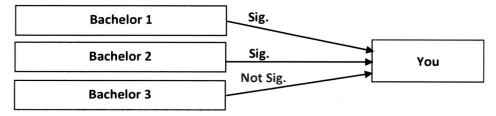

Step 6 (Data Analysis): Multivariate Analysis (Identify the Strongest Relationship).
In data analysis, the sixth step involves identifying the predictor variable that has the
strongest relationship with the dependent variable. This is typically done through a
multivariate model, such as a regression model.

For example, **Figure 2.13** presents the results of a multiple linear regression model,
which regressed the predictor variables *Income Level* and *Do You Live with a Dog or a
Cat?* upon the dependent variable *Happiness*. We can see that the effect size
relationship between *Income Level* and *Happiness* (effect size = .35) is higher than that
of the effect size between the predictor *Do You Live with a Dog or a Cat?* and
Happiness (effect size = .15). Thus, the strongest predictor of the dependent variable
Happiness would be the covariate variable *Income Level*.

Figure 2.13   The strongest predictor of *Happiness* (multiple linear regression)

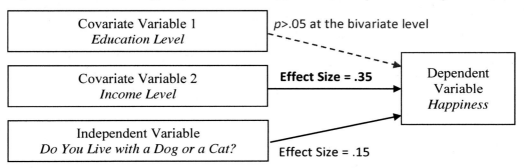

Step 6 (Dating Analysis): Identify the Strongest Relationship. In Dating Analysis, the
sixth step involves determining with whom you have the strongest relationship, out of
the pool of suitors, so you can meet your goal of selecting just one to go out with on a
date. For example, as presented in **Figure 2.14**, after considering the two bachelors
with whom you have a significant one on one relationship, you realize you enjoy going
out for Italian food and decide that you prefer bachelor 1, Bob.

Figure 2.14   Identifying the strongest relationship with a bachelor

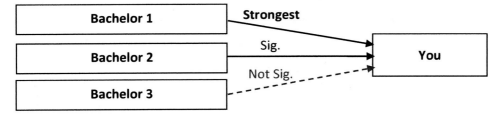

Step 7 (Data Analysis): Write-up & Report. Finally, the methods and results are reported in a professional style (e.g., AMA, APA).

Step 7 (Dating Analysis): Write-up & Report. The methods and results might also be reported in dating analysis, but instead of a professional format, the means is more likely to be personal (e.g., telling someone verbally, in a text, email, facebook, or journal entry).

In summary, at a purely conceptual level, *The 7 Steps of Data Analysis* model is no harder to understand than the process of online dating. The simplicity of this design is purposeful and is intended to make the model as accessible as possible.

2.5.3   A Clear Presentation of Ordered Steps

An essential feature of a professional model is *a clear presentation of the ordered steps* that must be taken to successfully implement that model. For example, **Figures 2.1, 2.2** and **2.3** present the clear order of steps that must be taken to conduct the power sweep football play. Obviously, if Coach Lombardi failed to present the correct order in which these steps must be followed, the team would have likely been confused. Furthermore, if the team was not aware of the order in which the steps should be applied, it is doubtful the play would have ever met with any success at all.

Like the football player and the football play, when one conducts a quantitative research study, he or she needs to know the correct order in which the necessary steps of data analysis must be applied. This is an important point as many students are regularly well educated on important facets of the data analysis model, but are rarely taught how the learned materials must be sequenced in order to conduct a comprehensive data analysis study. Subsequently, many learners and researchers end up in a similar situation as the football player who knows the steps of the power sweep play, but does not know how to arrange those steps in the correct order so that the play may be applied successfully.

One might wonder why students are rarely taught how the information they learn can be arranged into a stepwise progression that would facilitate a comprehensive data analysis study. It can be suggested that this might be a *byproduct* related to the need to present complex statistical procedures with an appropriate level of detail. For example, if a student needed to learn linear regression, he or she might take a semester long class where the details of the procedure are presented in depth. But, he or she is not likely to learn how linear regression is imbedded in the overall process of the data analysis study. Consequently, through this type of learning, students are often provided with a terrific level of detail regarding specific statistical procedures, but are not provided with a broad overview describing how these procedures are related and compose the overall process of data analysis.

Subsequently, the knowledge of how coursework learning must be ordered into specific sequential steps of data analysis has become a *missing link* type piece of information. Specifically, many students accumulate a great deal of knowledge regarding statistical procedures, but find themselves unable to apply that information. However, if students were aware of how conducting a quantitative research study breaks down into *The 7 Steps of Data Analysis* model, they would be aware of how to order and apply their coursework learning toward completing a comprehensive data analysis study. The confusion of students in applying coursework learning to conduct quantitative research is often exasperated by the fact that the knowledge that informs *The 7 Steps of Data Analysis* model is often taught out of order over the course of several semesters.

For example, let's say a student takes four classes (one per semester) that address specialized knowledge regarding the statistical procedures as presented in **Figure 2.15**. The figure describes that this student began his learning (semester 1 class) with a strong foundational course where univariate and bivariate statistical procedures were reviewed. The next semester the student studied linear regression (semester 2 class), followed by a course on power analysis (semester 3 class), and logistic regression analysis (semester 4 class). However, at the end of semester 4 when the student sits down to apply his learning toward conducting a formal quantitative study, he may find himself unsure of how to apply his learning in a comprehensive manner.

Figure 2.15   Four classes taken over four semesters on statistical procedures

SEMESTER 1 CLASS: Descriptive Statistics (e.g., mean, median, mode, standard deviation), T-test, Chi-Square, ANOVA, Correlation

SEMESTER 2 CLASS: Linear Regression Analysis

SEMESTER 3 CLASS: Power Analysis

SEMESTER 4 CLASS: Logistic Regression Analysis

However, if the student was aware of *The 7 Steps of Data Analysis* model, he would clearly recognize how to organize his learning and apply this information to complete a data analysis study. For example, as presented in **Figure. 2.16**, he would see that the class taken during semester three on Power Analysis actually addresses an aspect of Step 3 in our model Checks of Data Integrity. The foundational class taken in semester one on Descriptive Statistics, T-Test, ANOVA, Chi-Square, and Correlation addresses Step 4: Univariate Statistics and Step 5: Bivariate Statistics. The Linear and Logistic Regression courses taken during the second and fourth semesters (respectively), each address an aspect of Step 6: Multivariate Analysis.

Therefore, knowing the model of data analysis is an important key in applying the information toward conducting effective research. The aptitude to arrange and apply these steps can easily make the difference between coursework becoming a series of facts a student learned that may fade away in time, relative to a collection of valuable tools a student can apply in his or her professional field.

Figure. 2.16   How coursework is distributed within *The 7 Steps of Data Analysis*

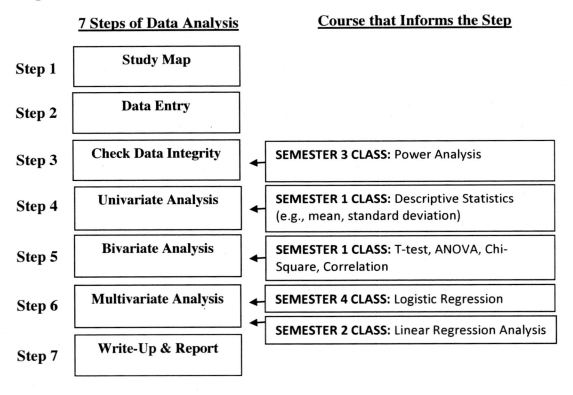

Enhancing Student Learning. It is also important to realize that knowledge of *The 7 Steps of Data Analysis* will not only benefit one in applying the statistical knowledge learned, but also in learning the materials as well. Specifically, learning would become more meaningful because students would be taught the materials with knowledge of the larger process in mind, or as it may be described, students would be taught the symphony, and not just the tuba.

To illustrate the symphony/tuba example, imagine you have never seen a musical instrument in your life, so you enroll in a music program to learn about the full range of instruments. Since you have never seen a musical instrument, you have never been exposed to a symphony. Thus, you are not only unfamiliar with instruments, but you are also <u>unfamiliar with how the instruments function together or perhaps even that the instruments do function together as a whole (i.e., the symphony)</u>.

So you arrive on your first day of school and go to your first scheduled class. The subject of this class is the tuba. Your first thought is "What is this rather cumbersome instrument?" You then think "It seems rather absurd and complicated." Then you shift into a sense of anxiety accompanied by statements such as "Why do I need to learn about this instrument? It seems rather pointless! How is it going to benefit my learning and life?" Worse than that, you have a bevy of students surrounding you saying the same thing and feeling the same way!

Now, let us consider the same scenario with a slight modification. In this instance, as a part of the music program, students were required to attend an <u>orientation</u> that introduced them to the categories of instruments. You learn how instruments are broken down into categories such as string, woodwind, brass, and percussion. Perhaps most notably, you are shown how all of these instruments function together to produce a marvelous symphony! You realize how each of these instruments combine to create a larger and more meaningful process as presented in **Figure 2.17**.

You may have noticed that the reaction many people have to statistics and research classes is similar to the initial reaction of the music student in the above scenario. In particular, many students in programs that have a required research component have never been exposed to the study of statistical research. Therefore, they are obviously not familiar with the fundamentals of the process. These students are often confused about the usefulness of learning statistical research or how the materials being learned can be applied.

Figure 2.17   Instruments & symphony: Seeing where the piece fits into the whole

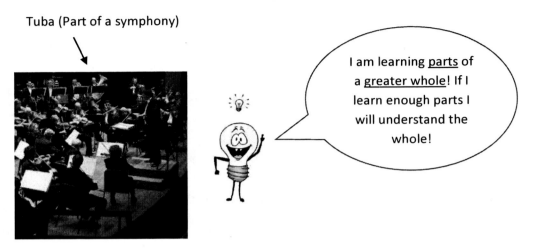

Tuba (Part of a symphony)

I am learning <u>parts</u> of a <u>greater whole</u>! If I learn enough parts I will understand the whole!

Pertaining to learning statistical research, a similar type of orientation could be presented, where a model of data analysis illustrates how statistical procedures function together to produce a marvelous data analysis study! Within this presentation, all seven steps within a data analysis study could be specified, along with a description of the statistical procedures incorporated within each step.

Subsequently, when students are introduced to a fundamental statistical procedure, such as the bivariate procedure known as the t-test, thoughts of anxiety and confusion might be replaced with their recognition that this test is fundamental to Step 5 (Bivariate analysis) within *The 7 Steps of Data Analysis*. Specifically, their first thoughts might reflect knowledge that this is an important tool used to identify if a significant one on one relationship exists between two variables, so the procedure is well worth learning.

Furthermore, as presented in **Figure 2.18**, learners may realize once they learn the essential procedures (e.g., the t-test) involved within Step 5, they may continue to learn the fundamentals of the other steps, until they develop an overall understanding of the entire process of data analysis. This perception that each small piece learned, contributes to understanding a greater whole, is perhaps the most meaningful point of view a student can have when learning statistical research.

The opposite of this condition would be a situation where students are unclear about what they are learning, why they are learning the materials, and perhaps only recognize that the materials seem difficult to them, which of course is precisely what we are trying to avoid.

Figure 2.18   Statistical test & data analysis: Seeing where the piece fits into the whole

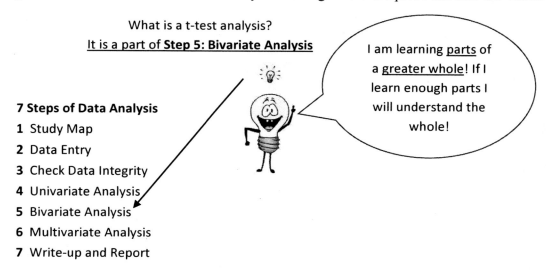

Therefore, a clear model to follow in performing statistical research may offer clear benefits in both learning and conducting data analysis.

2.6   A Game Plan for Success

To win a football game, you need a game plan. To conduct an effective data analysis study, you also need a game plan. In this section, *The 7 Steps of Data Analysis* model was considered as a game plan to learn, teach, implement, and conduct an effective data analysis study. The model was also scrutinized through examining if *The 7 Steps of Data Analysis* model reflects the characteristics of another near flawless model employed in another field (Coach Lombardi's power sweep football play) that produced dependable stellar results.

Through this discussion, we speculated that *The 7 Steps of Data Analysis* model appears to be a promising and effective tool for data analysis. However, there comes a point where discussion ends and doing begins, which is the point we have reached in this text. Next, after a brief introduction to necessary statistical concepts (**Part 3**), the remainder of the text will essentially apply *The 7 Steps of Data Analysis* model toward assessing and conducting actual data analysis studies. We are about to stop reviewing our play and begin running it!

PART 3

THE DATA ANALYSIS CONCEPTS YOU NEED TO KNOW

3.1 What Will This Section Tell Us?

This chapter will review several core concepts that you need to know before you approach a data analysis study. These concepts are so fundamental that they become second nature to the statistician, researcher, and data analyst. In fact, if you are not aware of them, much of the process of data analysis will not make sense to you. There are not many core concepts and most are simple, but knowledge of each is absolutely necessary.

3.2 Parametric and Non-Parametric Statistics

Statistical tests are often categorized as *parametric* and *non-parametric*. In this text, as well as in the world of quantitative research at large, parametric tests are largely applied as a means of data analysis. However, non-parametric tests are often useful resources to keep in mind. Parametric statistics are a group of tests that make certain *assumptions* (e.g., a normal distribution of scores) about the data used in analysis. Parametric tests presented in this text include the t-test, correlation, ANOVA, and linear (multiple) regression. If assumptions are met, these tests can produce accurate and precise estimates. However, if these assumptions are not met, results are likely to be misleading.

Non-Parametric statistics are a group of statistical tests that examine (typically through rank) the relationship between variables without making as many assumptions about the parameters and qualities of the data. A non-parametric test presented in this text is the chi-square analysis. Non-parametric tests are of use in many instances, such as when the dependent variable is an ordinal variable or a rank, as well as when the assumptions (e.g., no undue effect of outlier scores) of parametric testing are not met. Therefore, as we traverse statistical analysis, we will mention the non-parametric equivalent of many tests, when such tests are available.

3.3   Descriptive and Inferential Statistics

Statistical procedures are often categorized as descriptive or inferential. Before discussing the differences between descriptive and inferential statistics, it is helpful to review the concepts related to the *population* and *sample* as each pertains to quantitative research. A *population* is the total set of individuals, groups, objects, or events that the researcher is studying. For example, if a study focused upon the employment patterns of recent U.S. college graduates, the population might be defined as every college student who graduated within the past one year from any college across the United States.

A *sample* is a relatively small subset of people, objects, groups, or events selected from the population. This subset taken from the population is intended to *represent* the population in statistical analysis. For example, instead of surveying every recent college graduate in the United States (which would be impractical in term of cost and time constraints), a sample of this population would be surveyed and have their data analyzed.

An objective of such an analysis would then be to *generalize* the study findings generated from the study sample, to the larger population. Of course, there are a great many factors that impact how *representative* the sample is of the population, which effects how *generalizable* the findings are from the sample to the population. However, these factors often relate more to study design (e.g., a probability vs. non-probability sample) rather than statistical analysis. Therefore, since this text is focused upon applied data analysis rather than study design and related concepts, these concepts, as well as other important topics such as *probability distributions* will not be explored at length.

Inferential Statistics are the statistical procedures used to make predictions or inferences about a population based upon the observations made through the sample. In other words, the results of the *inferential* statistical analysis are used *to make inferences* about the larger population based upon the quantitative findings involving the sample. Within this text, inferential statistics are largely presented in **Step 5: Bivariate Analysis** (t-test, ANOVA, correlation, chi-square) and **Step 6: Multivariate Analysis** (multiple linear regression analysis and binary logistic regression analysis). *Descriptive statistics* (also described as *univariate statistics* in this text) are a group of statistical procedures used to describe the characteristics of a sample or population. Within this text, descriptive statistics are largely presented in **Step 4: Univariate Analysis**.

3.4   Level of Variable Measurement

Within this text, you will *repeatedly see that many of the essential decisions made in quantitative analysis are based upon considering the level of measurement* of a study variable(s) (also known as the *structure* of a variable). There are generally four levels of measurement regarding a study variable, which are *Nominal, Ordinal, Interval,* and *Ratio.* Within this section, we will review these four levels of variable measurement in detail, as well as present them in the context of composing categorical or continuous study variables. Specifically, in real world application, variable structure is often referred to as categorical (*Nominal* and *Ordinal* level measurement) or continuous (*Interval* and *Ratio* level measurement). Subsequently, in preparation of real world experience, in this section we will describe variable measurement as such.

Additionally, this section is also the first place we use the term *construct*. It is important to realize while this term is important in research, it is also a bit difficult to define. The word *construct* refers to the *idea(s)* associated with the label given to a variable or concept. It is as if the variable or concept is composed or *constructed* of several ideas and dimensions. For example, the variable *depression* is a construct that is associated with the dimensions of being unhappy, unmotivated, and so on. These dimensions are often measured through the identification of *indicators* that reflect the presence of the construct. For example, an interruption in sleep patterns is often an *indicator* of the *construct* depression.

3.4.1   Categorical study variables

Categorical variables place study participants (or any other *unit of observation*) into a *category*. This section describes the two general types of categorical variables, nominal and ordinal, as well as the definition of a dichotomous variable (a variable with a two category response). Although, the concept of a nominal or ordinal level of measurement is important, this detail is not reflected in the remainder of the text. Specifically, these variables are simply referred to as categorical and/or dichotomous.

Nominal variable. A nominal variable has two or more categories, *but cannot be ranked into an intrinsic order*, such as the variables gender (male/female) and racial identity (Caucasian/White, African American/Black, etc.).

Ordinal variable. An ordinal variable has two or more categories that can be ranked into a meaningful order. For example, if you were to rate a meal you might rank the order of categories as *poor, good, very good*, and *excellent*. However, there is *not an equal distance between categorical measurement points*. For example, there would not be a reason to believe there is an equal measurement distance between *good* to *very good*, in relation to *good* to *excellent*. In some research, *Ordinal* level variables are treated as continuous, but in this text, these variables are largely treated as categorical.

Dichotomous variable. Many times in research you will hear the term *dichotomous variable*. Therefore, it is important to realize that a dichotomous variable is a categorical variable (either nominal or ordinal level) that has a two category response, such as Yes or No, or good or bad.

3.4.2   Continuous study variables

Continuous variables reflect a greater or lesser degree of a *construct* as scores increase and decrease. This section presents the two general types of continuous variables, interval and ratio. However, this detail is not reflected in the remainder of the text, as these variables are simply referred to as continuous.

Interval variable. An interval variable measures the presence of a construct along a continuum with equal distances between measurement points, *but does not incorporate a meaningful true zero*. For example, an *IQ test* is an interval level measure of *general intelligence*. As IQ test scores rise, a greater presence of *general intelligence* is indicated. Additionally, there is an equal distance between measurement points where the distance between a score of 100 to 110 is 10 points, just like the difference between 120 and 130 is 10 points. However, there is not a meaningful true zero where a score of zero reflects a complete absence of general intelligence.

Ratio variable. A ratio variable measures the presence of a construct along a continuum with equal distances between measurement points and does *incorporate a meaningful true zero*. For example, measuring *weight* in pounds would be considered a ratio level measurement. As weight increases, a greater presence of the construct *weight* is indicated. Additionally, there is an equal distance between measurement points. For example, the distance between 100 and 110 is 10 pounds, just like the difference between 120 and 130 is 10 pounds. There is also a meaningful true zero score, as a weight of zero reflects a complete absence of the construct weight.

3.5   Single Versus Composite Item Scores

Another important issue in quantitative research concerns if study variables are measured using a single item or multiple items that may be combined into an overall composite score. For example, the study variables included in the sample studies presented in this text include variables based on single item score, such as *Education Level*, as well as composite item scores, such as *Happiness*. Each of these methods of measurement will be presented in this section.

3.5.1   The single item score

A single item score is a score that measures a study variable, which is based upon one item. For example, the sample studies included in this text each use the variable *Education Level*, which is based upon a single item score. Specifically, the variable *Education Level* is measured by one item from the study survey, which is presented in **Figure 3.1**. Each study participant is asked to select a response (a, b, or c) to the item *Highest Education Level Completed*. **The selected response to this <u>single item</u> will then reflect the score for the *Level of Education* of the study participant.**

Figure 3.1   Survey item 1: Highest Education Level Completed

1. Highest Education Level Completed:

   a. High School/GED
   b. College (4 Year Degree)
   c. Graduate School

3.5.2   The composite item score

A composite item score measures a study variable through combining multiple items to form a single score. For example, in sample study one the dependent variable *Happiness* is based upon combining five individual survey items (presented in **Figure 3.2**) to form a single composite score. The most common method of combining several items to compute a single composite score is to take the mean score of the multiple items (through summing the item scores and then dividing that summed number by the number of items). This will be the method applied to compute the composite *Happiness* score for each study participant. Generally, because a composite scale is composed of several items, the composite scale score is often considered a richer and more multidimensional measure of a construct relative to a single item score.

For example, consider how each of the five single scale items addresses a different dimension of the construct *Happiness*. Within the first two questions, study participants are not only asked about their level of feeling *happy* (question 1), but also their level of feeling *sad* (question 2). In other words, study participants are not only asked about the presence of feeling *happy*, but also about the presence of other emotions (feeling *sad*) that might suggest a lack of presence of *Happiness*. Through combining these multiple items, the composite score is thought to reflect the multiple dimensions represented by each item.

Figure 3.2   The five items composing the *Happiness* composite scale score

In the last 7 days I felt:					
	Strongly Disagree	**Disagree**	**Neither Agree or Disagree**	**Agree**	**Strongly Agree**
Happy	1	2	3	4	(5)
Sad	(1)	2	3	4	5
Good	1	2	3	(4)	5
Strong	1	2	3	4	(5)
Glad	1	2	3	(4)	5

3.6   Study Variable Type

In any quantitative study it is essential to identify the study variable type. For example, in the sample studies presented in this text, the study variable type includes independent, dependent, and covariate variables. Before statistical analysis can be approached, it is necessary to understand the roles of each type of variable in a study, as well as which specific study variables have been designated each role.

3.6.1   Independent variable

The **independent variable** (at times referred to as a **predictor variable** in this text) is the *variable of interest that explains the dependent variable* within the data analysis study. Specifically, the independent variable is tested toward identifying if the factor significantly predicts, causes, or impacts the dependent variable. For example, in the sample study one presented in this text, the factor *Do You Live with a Dog or a Cat?* is examined as an independent variable predicting the dependent variable *Happiness*.

3.6.2 Dependent variable

The **dependent variable** (also referred to as the **outcome variable** in this text) is the *variable being explained through the data analysis study*. The dependent variable is explained by the **predictor variables** (the independent and covariate study variables). For example, in sample study one presented in this text, the dependent variable *Happiness* is examined in relation to the independent variable *Do You Live with a Dog or a Cat?*, as well as the covariate variables *Education Level* and *Income Level*.

3.6.3 Covariate variable

The **covariate variable** (at times referred to as a **predictor variable** in this text and **control variables** in the field) is a *variable in a dataset that may not be the focus of the study, but must be included in analysis because the factor has a significant impact on the dependent or independent variables*. The covariate variable is frequently included in the data analysis study in an attempt to minimize or *control for* the effect of that variable on the relationship of interest between the independent and dependent variables. For example, in sample study one and two presented in this text, the factors *Education Level* and *Income Level* are included as covariate variables because we believe each has a significant influence on the dependent variable. Through including these covariate variables in analysis, the relationship between the independent variable and dependent variable (see **Figure 3.3**) can be seen more clearly, as the influence of each covariate variable upon the dependent variable is minimized or *controlled for*.

Figure 3.3   Study 1: Independent (IV) and dependent (DV) variables

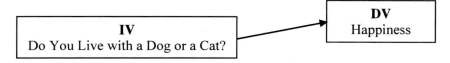

To better understand the motivation for including a covariate variable in this analysis first consider, that another researcher may say to us, the independent variable *Do You Live with a Dog or a Cat?* may have a moderate impact on *Happiness* in general. However, if you consider one's *Income Level*, living with a dog or cat would be almost inconsequential to one's level of *Happiness*. In order to address this posit, you would need to make *Income Level* (presented in **Figure 3.4**) a covariate variable, which would allow you to *control for* the influence of this factor, as you observe the relationship between the independent and dependent variables. If the relationship between the independent and dependent variables is significant while controlling for the covariate variable *Income Level*, the researcher's posit would **not** be supported.

Figure 3.4   Study 1: The covariate variable (CV) *Income Level*

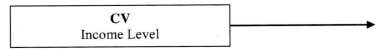

If we look deeper at his posit, the hypothetical researcher is suggesting that *Income Level* is such a powerful predictor of the dependent variable *Happiness*, that the covariate variable *Income Level* in a sense *washes away* the impact of the independent variable *Do You Live with a Dog or a Cat?* on the dependent variable *Happiness*. This effect is illustrated in **Figure 3.5** where in the top of the tidal wave, *Income Level* is such an overarching and powerful predictor that it washes away the effect of the independent variable *Do You Live with a Dog or a Cat?* on the dependent variable *Happiness* like a tidal wave.

Figure 3.5   The covariate variable *Income Level* washing away the effect of the independent variable *Do You Live with a Dog or Cat?* on the Dependent Variable *Happiness*

However, the opposite might be true, where the independent variable *Do You Live with a Dog or a Cat?* is such a powerful predictor that it would wash away the effects of the covariate variable *Income Level* on the dependent variable *Happiness*. This effect is illustrated in **Figure 3.6** where in the top of the tidal wave, the independent variable *Do You Live with a Dog or a Cat?* is such an overarching and powerful predictor that it washes away the effect of *Income Level* on the dependent variable *Happiness*.

The more powerful predictor can be identified through including both within a multivariate statistical procedure (e.g., multiple linear regression) where the **joint impact** of both the covariate and independent variables is observed upon the dependent variable *Happiness* **together** as presented in **Figure 3.7**.

Figure 3.6   The independent variable *Do You Live with a Dog or a Cat?* washing away the effect of the covariate variable *Income Level* on the dependent variable *Happiness*

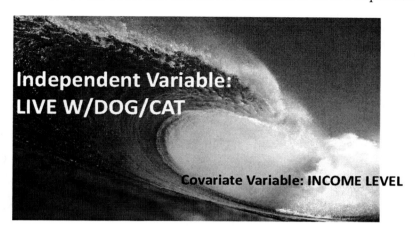

Figure 3.7   The joint effect of the covariate (CV) and independent (IV) variables

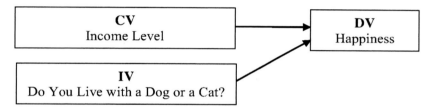

Through this method of including two predictor variables within a multivariate regression model, you are *controlling for* the effect of the covariate variable in the model, while observing the relationship between the independent and dependent variables. If within the regression model, the relationship between the independent variable *Do You Live with a Dog or a Cat?* and dependent variable *Happiness* is statistically significant, while controlling for the covariate variable *Income Level*, it can be said that the independent variable predicts the dependent variable *over and above* the covariate variable. Thus, the challenge presented by the hypothetical researcher would have been addressed via analysis and unsupported by the data. When you think about it, it is pretty amazing to be able to test the impact of two factors (e.g., predictor variables) on an outcome (e.g., the dependent variable) to determine which predictor has the strongest impact on the dependent variable.

Teasing Out the Effect of a Covariate Variable: "It's a Wonderful Life" as an Example. When we discuss covariate variables, many times we are discussing what is known in statistics as *teasing out* the effect(s) of one or more factors on the variable relationship of interest (i.e., between the independent variable and dependent variable).

Teasing out the effects of a factor upon a relationship of interest is actually a common theme in literature and film. For example, consider the classic film *It's a Wonderful Life* starring Jimmy Stewart.

In the movie, an angel (Clarence) shows the central character (George Bailey) what the lives of people in his community of Bedford Falls would have been like if George had not been born. What the angel was essentially doing was *teasing out* or *controlling for* the effect of George's actions upon the relationship between the people of Bedford Falls and their life outcomes. One of the most serious life outcomes presented in the movie, concerned the pharmacist in town for whom George worked as a teenager. While George worked at the pharmacy, the pharmacist mixed an incorrect and lethal combination of medicines to fill a prescription that would be given to a young girl. George pointed this mistake out and the pharmacist fixed the compound, which saved the life of the girl who was to receive the prescribed medicine.

We could consider that the *life outcome* in this instance would be if the *life of the girl with the prescribed medications* was *saved* or *not saved*. The relationship between the people of Bedford Falls (the pharmacist) with the effect of George's actions on this life outcome is presented in **Figure 3.8**. This figure illustrates, when the effect of George is present, the result regarding this life outcome is that the *life of the girl with the prescribed medications* was *saved*.

Figure 3.8   Life Outcome (LO) WITH the effect of George

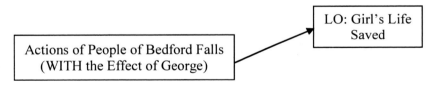

However, later in the film the angel *teases out* the effect of George's actions from this relationship to illustrate how the life outcome would have differed if George had never been born. Specifically, Clarence the angel showed George the scenario where George was not present to point out the pharmacist's mistake, which resulted in the young girl being given the lethal mixture of drugs and her life not being saved. Thus, Clarence showed George that once he *teased out* the effect of his actions from the relationship between the actions of the people of Bedford Falls and their life outcomes, their life outcomes changed dramatically. **Figure 3.9** presents the relationship between the actions of the people of Bedford Falls without the effect of George's actions on their life outcomes. This figure describes when the effect of George is not present there is a negative life outcome where the *life of the girl with the prescribed medication was not saved*.

Figure 3.9   Life Outcome (LO) WITHOUT the effect of George

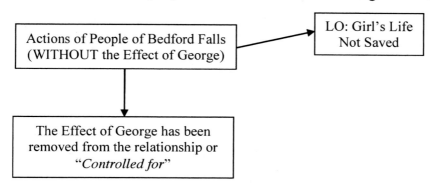

Similarly in statistics, when we include a covariate variable many times we are in a sense teasing out the effect of that covariate variable from the relationship between the independent and dependent variables, just as Clarence teased out the effect of George's actions. For example, suppose we wanted to examine the relationship between the *Number of Cigarettes Smoked* (the Independent Variable) and the *Number of Aggressive Behaviors* (the Dependent Variable) displayed by study participants as presented in the **Figure 3.10**.

Figure 3.10   The independent and dependent variable relationship

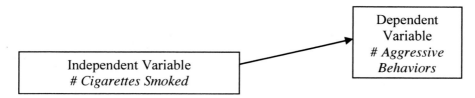

Furthermore, suppose you read that aggressive behavior is more prominent in males than females, so you decide to control for or tease out the effect of gender from this relationship. You could do this by adding *gender* as a covariate variable in this analysis as presented in the **Figure 3.11**. Through structuring the analysis in this manner you would have teased out the effect of gender and better isolated the relationship of interest between the independent variable and dependent variable (*# cigarettes smoked* and *# aggressive behaviors*, respectively).

We will review the *multivariate statistical procedures* needed to test these relationships in **Sections 4.7.2** (Conducting Multiple Linear Regression Analysis) and **5.7.2** (Conducting Binary Logistic Regression).

Figure 3.11 Controlling for the covariate variable *Gender*

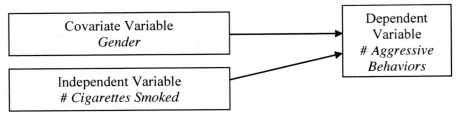

In end, while the covariate variable(s) may not be the focus of the study, these variables are essential to the study. These important factors are instrumental in accounting for the complex set of influences on the other study variables, which is vital to revealing the true relationship between the variables that are the focus of the study.

3.7   The Clarity of the Study Hypothesis

A hypothesis is a predictive statement regarding the relationship between study variables. The hypothesis is the central figure in a quantitative study. Specifically, the study hypothesis is like the sun within the universe known as the data analysis study, around which the planets known as the steps of data analysis revolve. Quite appropriately, the essentials of the study hypothesis are regularly presented in statistics textbooks. For example, the hypothesis(es) is typically presented in terms of:

The Alternative and Null Hypothesis. The alternative hypothesis (also called the research hypothesis) and null hypothesis are the two rival hypotheses, one of which will be supported through statistical testing. The **alternative hypothesis (research hypothesis)** is the relationship between variables that the researcher believes exists and is testing for (e.g., *There is a significant relationship between Living with a Dog or a Cat and Happiness*). The **null hypothesis** is the default position that states that there is no relationship between the variables under examination (e.g., *There is not a significant relationship between Living with a Dog or a Cat and Happiness*).

Statistical testing ultimately results in _rejecting_ either the alternative (research) or null hypothesis based upon the results of statistical testing. For example, if the statistical testing indicated that *Living with a Dog or a Cat* and *Happiness* are significantly related ($p<.05$), the null hypothesis will be rejected and the alternative hypothesis (research hypothesis) would be supported. However, if the statistical testing indicated that *Living with a Dog or a Cat* and *Happiness* are not significantly related ($p>.05$), the null hypothesis will be supported and the alternative hypothesis (research hypothesis) would be rejected.

Type I Error and Type II Error. Also conceptually relevant to the hypothesis is the concept of **type I error** and **type II error**. **Type I error** involves a rejection of a true null hypothesis. Specifically, type I error leads one to conclude that a significant relationship between variables exists, when in reality it does not. For example, a type I error would occur when analysis indicates a significant relationship exists between *Living with a Dog or a Cat* and *Happiness*, when the relationship <u>does not</u> exist. A **type II error** is the failure to reject a false null hypothesis. Type II error would occur when analysis indicates a significant relationship does not exist between *Living with a Dog or a Cat* and *Happiness*, when the relationship <u>does</u> exist.

Directional and Non-Directional. The hypothesis is also described as being directional or non-directional. A directional hypothesis posits how variables will be related, such as the hypothesis stating *People Living with a Dog will have higher levels of Happiness relative to those Living with a Cat*. A non-directional hypothesis posits a relationship exists, but does not specify how the variables will be related, such as the hypothesis stating *People Living with a Dog will have significantly different levels of Happiness relative to those Living with a Cat*.

The concept of the directional hypothesis relates to the **one-tailed test** and **two-tailed test**, which is available in certain statistical procedures (e.g., correlation, t-test). Specifically, some tests provide an option to select a one-tailed test or two-tailed test, which are alternative ways of computing the statistical significance ($p<.05$). The **one-tailed test** is said to be appropriate when a directional hypothesis is considered, while a **two-tailed test** is said to be appropriate when a hypothesis is non-directional.

However, you may notice within the quantitative research literature (as well as the statistical analysis conducted in this text) the two tailed test is applied consistently, even when a hypothesis is directional. This is likely related to the fact that the two-tailed test is a more rigorous test, which may suggest a more stable relationship between variables. In other words, a one-tailed test indicates a statistically significant relationship more easily, so we might favor the more stringent two-tailed test. Thus, even though the hypotheses are directional in sample studies one and two, analysis will only employ two-tailed tests.

However, beyond the fundamental concepts related to the hypothesis, lies the essential consideration of hypothesis *clarity*. *Clarity* is an indispensable virtue related to the hypothesis. Therefore, we will now discuss several elements related to hypothesis clarity, as well as a test to examine if the hypothesis is clear in terms of being succinct.

3.7.1 Succinct

A potentially useful test that can be applied to examine if the hypothesis is succinct is the *first half/second half test*. I noticed when I began writing hypotheses they were so long that by the time I read the second half of my hypothesis, I had forgotten what the first half said. I realized that was a bad sign, so a regular test of mine became *reading the hypothesis aloud and observing if the hypothesis was held in my mind as one thought without rereading it*. Chances are if you cannot do this, your hypothesis can be shortened and made more succinct.

3.7.2 Clarity

The hypothesis is to an empirical study, what the bull's-eye in the center of a target is to an archer. The archer needs that bull's-eye to focus and guide his or her actions to achieve profound success. All actions, thoughts, and considerations revolve around hitting that bull's-eye. The more clearly the archer can see the bull's-eye, the better are his or her chances of success. If the archer's view of the bull's-eye is cloudy, the better the chances the target will be missed. Likewise, for a quantitative researcher, all the actions, thoughts, and considerations involved in the data analysis aspect of a study revolve around effectively answering the study hypothesis(es). Furthermore, the more clearly the researcher can see the hypothesis, the more likely he or she will be able to effectively indicate if the hypothesis is supported or not supported. If the researcher's view of the hypothesis is unclear, the better the chances the hypothesis will not be answered properly. Thus, please note the following three points regarding hypothesis clarity to maximize the chances of study success.

Point 1: Present one relationship at a time. The first step in making the hypothesis clear is to first identify the constructs (e.g., independent and dependent variables) to be included within the hypothesis and be sure only <u>one relationship among variables</u> is being hypothesized. Many times you will see several variable relationships being hypothesized and then presented as a single hypothesis, such as:

People living with dogs are <u>happier</u>, <u>healthier</u>, and <u>more optimistic</u> than people living with cats (we will call this the original longer hypothesis).

This hypothesis is actually three shorter hypotheses, namely:

Hypothesis 1: People living with dogs are <u>happier</u> than people living with cats.

Hypothesis 2: People living with dogs are <u>healthier</u> than people living with cats.

Hypothesis 3: People living with dogs are <u>more optimistic</u> than people living with cats.

(we will call these the three shorter hypotheses)

One of the reasons the original longer hypothesis needs to be broken up into the three shorter hypotheses is because of the need to definitively state if a hypothesis is supported or not supported. If several relationships are presented in a single hypothesis, then some may be supported and some may not be supported by the data, which would preclude our ability to state if the hypothesis is definitively supported or not. Using the original longer hypothesis as an example, the findings could have indicated that:

*People living with dogs **ARE** <u>Happier</u> than people living with cats.*

*People living with dogs **ARE** <u>More Optimistic</u> than people living with cats.*

*People living with dogs **ARE NOT** <u>Healthier</u> than people living with cats.*

Subsequently, you can't definitively state if the original longer hypothesis was supported or not supported as parts of the hypothesis were supported (ARE *Happier*, ARE *More Optimistic*), while other parts were not (ARE NOT *Healthier*). However, if the longer original hypothesis was broken into the three shorter hypotheses you could definitively state if each was or was not supported using the same findings as such:

Hypothesis 1: People living with dogs are <u>happier</u> than people living with cats.
Hypothesis 1 was supported by the data.

Hypothesis 2: People living with dogs are <u>healthier</u> than people living with cats.
Hypothesis 2 was supported by the data.

Hypothesis 3: People living with dogs are <u>more optimistic</u> than people living with cats.
Hypothesis 3 was not supported by the data.

To recap, the first suggestion regarding making your hypothesis clear is to present *only* one relationship at a time.

Point 2: Making the cause and effect relationship clear. In sample study one we are hypothesizing a direct cause and effect relationship. Specifically, we are hypothesizing that *Happiness* is directly impacted by the variable *Do You Live with a Dog or a Cat?* When you are presenting a hypothesis that addresses a direct relationship, it is often useful to see if it passes the *"If A then B"* test to be sure you are clearly identifying the cause and effect relationship. For example, regarding sample study one, our hypothesis would state:

***If** a person lives with a dog, **then** he or she will have a higher level of Happiness relative to a person that lives with a cat.*

Of course the hypothesis does not need to be presented in this format. However, it is usually a good idea to check that the hypothesis *can* be phrased in this way as a means of examining if a clear cause and effect relationship is being presented.

Point 3: Specify the variables being examined. Perhaps the most important factor in making your hypothesis clear is to specify the variables being examined as clearly as possible. For example, the study hypothesis for sample study one would be much less clear if we were more general in stating the study variables as:

The Type of Pet Owned will be significantly related to levels of Happiness.

Notice, although it may seem like a clear hypothesis is being presented, the statement is so general that it does not describe an exact testable relationship. However, if we replace the term *type of pet* owned with the more specific *lives with a dog* or a *cat* (as is done in the actual hypothesis) the relationship we are examining is much clearer:

People who Live with a Dog have a higher level of Happiness relative to People that Live with a Cat.

Thus, the clearer the terms we use to describe the variables being studied, the more powerful the hypothesis will be.

3.8   Three Dimensions of a Relationship

The goal of most data analysis studies is to examine the relationship between study variables. However, before such a relationship can be examined, a researcher must be aware of the three dimensions of any variable relationship that must be considered, which are *significance, directionality*, and *magnitude*. These three dimensions of a variable relationship are the same three aspects that would effectively describe a human relationship. To illustrate this similarity, let's compare how these three aspects of a variable relationship would be examined in a human relationship, such as a friend going out on a first date.

3.8.1   Aspect 1: The significance of the relationship

The first consideration of a relationship is significance. If a relationship is not significant, you can essentially dismiss any other relationship aspects. This section presents how significance applies to relationships between both humans and variables.

The Friend and the Dater: Suppose you saw a friend at work who went out on a first date the night before (with Bob from *d* Harmony mentioned in **Part 2**). Let's also assume you might be curious about her experience, so you ask your friend about her feelings regarding the date. There are generally two reactions you might receive:

First, your friend might shrug her shoulders and make a slightly unpleasant contorted face and say "Eh." This is typically a sign that there is no significant feeling toward her dating companion and usually signifies that the relationship is not significant. The discussion would most likely end there, since there seems to be nothing additional to discuss. Alternately, your friend might say something like "Oh, I have some feelings about that person" (Bob). This response reflects that this might be a significant relationship. As a result, you would likely examine other aspects of the relationship (i.e., directionality, magnitude). Please note, in our example, the indicators of a significant relationship are:

"Eh" = Not a Significant Relationship

vs.

"Oh, I have some feelings about that person" = Significant Relationship

The Statistics: In testing the relationship between variables, significance, or more specifically, statistical significance is expressed through the level of probability (a.k.a., p-value), which reflects how likely the relationship presented is a result of chance. Although the level of probability may vary, the typical scientific standard is that probability must be less than 5 times in 100 (i.e., $p<.05$) to achieve statistical significance.

Similar to our example regarding human relationships, if the statistical relationship is not significant, one would not consider the other two aspects (i.e., directionality, magnitude) of the relationship between variables. If the relationship does not achieve statistical significance it is assumed to be by chance, and therefore any other aspects of the relationship are not credible. Thus, the indicators of significance in statistical testing are:

$p>.05$ = Not a Significant Relationship

vs.

$p<.05$ = Significant Relationship

It is important to realize, that reporting the level of probability, i.e., is probability greater than .05 ($p>.05$) or equal to/less than .05 ($p<.05$), is one of the most focused upon activities in inferential testing. Many times people realize that p (probability) must be at or below .05 for a finding to be statistically significant, but are not sure what that means. As mentioned above, setting the probability level at .05 means that 5 times out of 100 a significant finding would be by chance, but what does that mean?

Essentially, when performing a statistical test, we are asking the statistical procedure to tell us if a significant relationship exists between variables (e.g., Does a significant relationship exist between levels of _Happiness_ among people that live with a _dog_ or a _cat_?). However, at the same time, we realize that sometimes the statistical procedure will in a way lie to us, where the test will indicate a significant relationship exists, when it actually does not. Therefore, we set probability to specify how many times we are willing to be told a significant relationship exists when actually the finding is a result of chance, with the standard being 5 times out of 100 ($p<.05$).

It is rather like, if you knew someone that gave you tips for race horses to bet on. Suppose this person was known to provide you with both winning tips, as well as losing tips. Well, these winning and losing tips could be expressed as a percentage reflecting how many times the tips were accurate, such as 50% of the time the tips given by this person were accurate winners. Realistically, this person would need to give you accurate information a high percentage of the time before you would consider believing the information provided.

In essence, there would be a **cutoff point** for accurate information that must be met before you would listen to him. For example, if he gave you accurate tips that turned out to be winners only 40% of the time (i.e., inaccurate 60% of the time), you might not act on his tips since over half of the time he was leading you astray. However, if 80% of his tips were accurate (i.e., 20% were inaccurate), you might be more willing to listen to him. When we set the probability level at .05, it is as if we are saying that we will only believe these race horse tips if they are accurate 95% of the time (inaccurate 5% of the time). We are again **selecting a cutoff point** for the rate for which we will accept accurate versus inaccurate information.

However, regarding statistics, we are not referring to horse race tips, but how often we are willing to accept inaccurate information regarding variable relationships from a statistical test. Our **cutoff point** is .05 ($p<.05$), which means 95% of the time the statistical test will supply *accurate* information about relationships between variables (the equivalent of accurate race horse tips), but 5% (i.e., .05) of the time we will be told *inaccurate* information about relationships between variables (the equivalent of inaccurate race horse tips) where findings appear significant by *chance*.

That may lead you to ask, what does it mean when significant relationships occur by *chance*? You could think of a *chance* occurrence as a development that gives the appearance of a certain situation, but is actually not that way and is something else (see any episode of the sitcom *Three's Company* as a reference). For example, let's say you did not like pizza. However, one day at work a client gave you his slice of pizza to hold for him as he signed a million dollar contract for you. Right at that point a co-worker saw you holding that slice of pizza and asked you "Oh, you like pizza?" And you say "No, I just happen to be holding this slice *by chance* when you saw me." Note, it appeared that there was a significant relationship (between you and pizza). However, there was not a significant relationship, it only *appeared* that way out of a chance occurrence (you happened to be holding a slice of pizza).

In statistics, a chance occurrence does not look like you and a slice of pizza. The chance occurrence would more likely be a statistically significant relationship between variables on the computer that is not really significant in the real world. For example, consider our research question: Does a significant relationship exist between levels of *Happiness* among people that <u>live with a *dog* or a *cat*</u>? Suppose the findings indicated that people living with *dogs* had twice the level of *Happiness* relative to people living with *cats*, which our statistical test indicated was a statistically significant relationship. This could be expressed as:

1) The average *Happiness* score for people living with dogs is **4.0**.

2) The average *Happiness* score for people living with cats is **2.0**.

3) The statistical test indicated the relationship between *Happiness* scores and living with a dog or a cat (i.e., the difference in the two average *Happiness* scores) is statistically **significant** at the **.05 ($p<.05$)** level.

In terms of probability being set at .05, we are saying that 95% of the time the above finding will be an accurate significant relationship that actually exists between variables. However, 5% of the time this finding will not be an accurate significant relationship that exists between variables, but will be the result of *chance*, where things sort of fall in place in a way that makes a relationship look significant when it is really not. Notice, the **statistical findings stay absolutely the same**, but conceptually in our minds we acknowledge that 5% of the time they will not be accurate, but will occur by chance. Thus, it is important to keep in mind when a statistical test indicates that a relationship between variables is significant, **we should not accept that this is absolutely the case**. In explaining the significant relationship between variables, always remember that the results may have nothing to do with the variables, but might be a function of the *chance* occurrence inherent in statistical testing.

Keep in mind, when probability is set at .05, 5 times out of 100 significant findings will occur by chance, which could also be expressed as 1 out of 20 significant findings will occur by chance. Therefore, if you conduct 20 statistical tests (or around that number) and identify one statistically significant finding from that pool of tests, it is likely that this one significant finding is the finding that has occurred by chance. In statistics, you might hear the term *fishing expedition*, which often refers to a researcher conducting many statistical tests to *fish* for a statistically significant finding. This often implies that the statistically significant finding(s) *fished for* is likely to be the chance finding that will occur when enough statistical tests are conducted.

3.8.2 Aspect 2: The directionality of the relationship

Furthermore, in the case of statistics and people, after you establish if a relationship is significant or not significant, you might then wonder if the relationship is significantly positive or negative. The question regarding if a relationship is positive or negative refers to the *directionality* of a relationship. In this section, we will discuss how directionality applies to relationships between both humans and variables.

The Friend and the Dater: Once your friend indicates the relationship is significant through stating "Oh, I have some feelings about that person," your next question might be a query regarding if the significant feelings are positive or negative. Specifically, one could have significant feelings, but these feelings might be significantly positive emotions or negative emotions. This is a question that reflects the directionality of the relationship. This might be indicated by your dating friend through a statement such as:

"I like him" (Bob) = Positive Relationship

vs.

"I do not like him" (Bob) = Negative Relationship

The Statistics: When we are testing the relationship between variables, once we have identified a relationship is significant, we would then want to observe the second relationship aspect, which is directionality. Directionality here would indicate a negative or positive relationship between variables. A **positive relationship** between two variables in statistics could mean, as the first variable increases in score, the second variable increases in score. For example, a positive relationship between the two variables *Income Level* and *Happiness* would suggest as *Income Level* increases, *Happiness* scores also increase. A **negative relationship** between two variables in statistics could mean as the first variable increases in score, the second variable decreases in score. For example, a negative relationship between the two variables *Income Level* and *Happiness* would suggest as *Income Level* increases, *Happiness* scores decrease.

Typically, in inferential statistical testing, a negative relationship is indicated by the presence of a minus (-) sign in front of the statistical coefficient. A positive relationship is indicated if there is not a minus sign (no -) in front of the statistical coefficient. The statistical coefficient is the statistic produced in each statistical test, such as the *Pearson's r* coefficient produced in a bivariate correlation. This will be reviewed in greater detail, but as a short introductory example, if the correlation yields a correlation coefficient of $r = -.35$ (the number .35 selected here is arbitrary) the relationship is negative (notice the minus sign), but a correlation coefficient of $r = .35$ (no minus sign), indicates a positive relationship. Thus, in statistical testing the indicators of directionality are:

Minus (-) Sign = Negative Relationship

vs.

No Minus (no -) Sign = Positive Relationship

3.8.3   Aspect 3: The magnitude of the relationship

Finally, once you find out if a relationship is significant, as well as positive or negative, you might then wonder how strong the relationship is, which is a question of magnitude. In this section we will discuss how magnitude is equally relevant to human and variable relationships.

The Friend and the Dater: Once your friend indicates that the relationship is significant, and perhaps positive, many times you might ask them "I know you *like* Bob, but do you *really* like him?" This query is typically the qualifying question to determine if your friend likes the person enough to go out on another date. This question reflects a measurement of magnitude regarding how strong that relationship might be. It is possible to have a relationship that is significant (whether positive or negative), but not terribly impactful.

The classic example in dating is when someone says "I do like him, but *something* is just not there." What is being said is that the relationship might be both significant and positive, but the magnitude is (for the purposes of dating) weak. Subsequently, in our dating scenario, magnitude would be indicated by your friend telling you:

Do you *really* like him: "Yes" = Strong Magnitude

vs.

Do you *really* like him: "No" = Weak Magnitude

Statistics: In inferential statistical procedures the magnitude of a relationship can be indicated several ways, including the calculation of an effect size. Essentially, an effect size reflects how impactful one variable is upon another, thereby estimating the magnitude of a relationship between variables. For example, if we consider what factors impact a variable such as *Happiness*, we could imagine that there are things that might be very impactful on *Happiness*, such as health or income, as well as things that are less impactful, such as having a bad hair day.

The Small, Medium, and Large Effect Size. In statistical testing, the magnitude of an effect size is categorized as being small, medium, or large. Before we discuss this concept statistically, it may be useful to examine how the concept of small, medium, and large effect sizes relate to things that occur in our everyday lives.

For example, let's say you have an everyday outcome in your life such as *How Much You Enjoy Your Dining Experience*. Specifically, let's say you are going to rate the level of enjoyment of a three course (**appetizer, entrée,** and **dessert**) dining experience. In this case, it is likely that each course might have **a differential impact** on your rating of how enjoyable the dining experience is. In other words, each course would play a greater or lesser role in **predicting (Predictors:** *appetizer, entrée,* and *dessert*) the outcome (**Outcome:** *How Much You Enjoy Your Dining Experience*). For example, in this scenario let's assume you have the following feelings regarding each of these three courses in general:

Course #1 The Appetizer: The *appetizer* is your least favorite course. If the appetizer is good or bad, it does not impact you much, as you hold the appetizer in rather low esteem. Therefore, you may describe that the appetizer has a rather *small* size effect on how much you enjoy your dining experience.

Course #2 The Entrée: The *entrée* is your favorite course. In fact, the entrée by itself will practically determine if you do or do not enjoy your meal. Therefore, the entrée has a *large* effect on how much you enjoy your dining experience.

Course #3 The Dessert: Lastly, while the appetizer has only a small effect on your dining experience, you do enjoy *dessert*. However, you do not consider dessert as important as the entrée. Therefore, you might consider dessert to have a *medium* size effect on how much you enjoy your dining experience.

Figure 3.12 illustrates how the concept of a small, medium, and large effect size might be evidenced by each of these three courses upon the outcome *How Much You Enjoy Your Dining Experience*. For example, you may be able to conceive of how your least favorite course, the appetizer, would have a small effect size on your dining experience, while the favorite course, the entrée, would evidence a large effect size on *How Much You Enjoy Your Dining Experience*. Likewise, you may be able to imagine how the dessert, which is more important than the appetizer to you, but less important than the entrée, might have a medium effect size on *How Much You Enjoy Your Dining Experience*.

Figure   3.12   Small, medium, and large effect sizes among predictors of an outcome

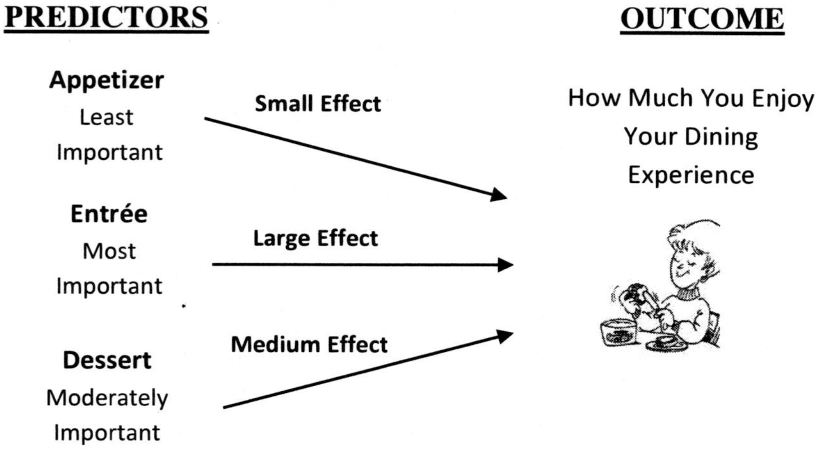

PREDICTORS   **OUTCOME**

Appetizer   *Small Effect*   How Much You Enjoy
Least   Your Dining
Important   Experience

Entrée   *Large Effect*
Most
Important

Dessert   *Medium Effect*
Moderately
Important

How Can We Indicate a Small, Medium, and Large Effect Size Statistically? Now
that we have discussed small, medium, and large effect sizes, we might wonder how
each can be reflected statistically. As indicated in **Figure 3.13**, these categories of
effect size are reflected by clear numerical values. Please note, the actual raw numerical
values reflecting a small, medium, and large effect size often varies by the statistical
test used (listed in the left hand column). However, all effect size numbers can be
categorized as small, medium, or large. For example, the raw numbers for an r effect
size describe that a small, medium, and large effect size would be indicated by the
values 0.01, 0.09, and 0.25, respectively. However, the raw numbers reflecting a small,
medium, and large effect for the eta squared effect size are 0.01, 0.06, and 0.14,
respectively. Thus, keep in mind while the raw numbers of effect sizes may not be
comparable, the categories of small, medium, and large effect sizes will be. As we
conduct the statistical procedures listed in the column to the far right in this Table, we
will review how to produce and interpret many of these effect sizes.

Thus, an indicator of the magnitude of the relationship between variables could be
described as:

Small Effect Size,

Medium Effect Size,

or

Large Effect Size

In summation. Whenever you conduct an inferential statistical test, you should look at
the statistical output for indicators of the three aspects of a variable relationship, which

are significance, directionality, and magnitude. As a mnemonic to remember these three aspects, try a term that reflects something like in a *relationship* you must *Savor Daily Miracles* (SDM). An example of a daily miracle in a relationship would be when someone close to you pays you a spontaneous compliment. When things like this happen, it is important to savor these daily miracles. Therefore, when you think of a relationship, think of SDM, S=Significance, D=Directionality, and M=Magnitude.

Figure 3.13   Small, medium, and large effect sizes by statistical tests

Effect Size	Small	Medium	Large	A Test Where You Will See this Effect Size Used
R	0.10	0.30	0.50	Pearson *r* Correlation
r	0.01	0.09	0.25	Pearson *r* Correlation
Eta-squared	0.01	0.06	0.14	ANOVA
R	0.01	0.06	0.14	Linear/Multiple Regression
Cohen's *d*	0.20	0.50	0.80	T-Test
Phi	0.10	0.30	0.50	Chi Square (2X2)
Cramer's V	0.10	0.30	0.50	Chi Square >(2X2)
Cohen's f²	0.02	0.15	0.35	Linear/Multiple Regression
Odds Ratio	1.44	2.47	4.25	Logistic Regression

3.9   The Russian Doll Effect in Analysis

As you perform each necessary step and procedure within a data analysis study, you will notice that each task incorporates many subtasks, which also incorporate many smaller tasks. I like to call this tendency the "Russian Doll" effect of data analysis. Specifically, as presented in **Figure 3.14**, a set of Russian Dolls typically contains a series of dolls where the smaller dolls fit into the larger dolls until only one is visible.

Figure 3.14   A Classic Assembly of Russian Dolls

A task in a data analysis study is very much like this. Specifically, in data analysis there is often one large task visible, but when that large task is broached there are many smaller tasks within. For example, within **Figure 3.14** the largest doll reflects that the largest overall construct in data analysis is **The 7 Steps Data Analysis** model. If we consider **Step 3: Checks of Data Integrity** of the overall model, the second largest doll reflects within this step, we will find four checks of data integrity that need to be addressed. If we consider the second check of data integrity, **2) Test Assumption**, the third doll reflects there would be several assumptions to examine, including **1) Normality**. The final doll reflects that within the check of **1) Normality** there are several other considerations (i.e., Skewness, Kurtosis, Outlier scores). Just like the Russian Dolls, every step in data analysis breaks down into a series of smaller pieces.

Figure 3.14   The Russian Doll Effect for **Step 3: Checks of Data Integrity**

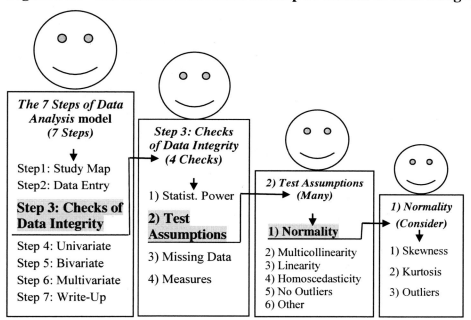

3.10   Haste or Paste: Save the Syntax

Throughout this text, the instructions for each statistical test and procedure end with "Click **OK** or **Paste**." If you choose to click **OK** the statistical test or procedure will be implemented immediately. If you choose to click **Paste**, the SPSS software will produce the syntax necessary to implement the procedure, which can then be saved as a syntax file. Once the syntax is in this file, all you need to do is *highlight the syntax and run the test by clicking the run icon*, which is the large green button in the top center of the screen. It may be a bit more expedient to click **OK**, but I would urge you to click **Paste**, so you can produce and save your syntax. This is useful on a number of levels, including producing an electronic paper trail verifying the procedures were conducted properly, as well as the fact that you can repeat the procedures through opening and running the syntax instead of repeating the entire analysis.

PART 4

A QUANTITATIVE STUDY WITH A CONTINUOUS DEPENDENT VARIABLE

4.1   What Will This Section Tell Us?

In Part 4, through sample study one, we will review how to conduct a quantitative study that examines a continuous dependent variable. Specifically, we will examine how scores for the continuous dependent variable *Happiness* (measured using a composite scale score of 1-5) are impacted by several predictor variables. Each predictor variable will incorporate a different level of measurement. Specifically, the independent variable *Do You Live with a Dog or a Cat?* is dichotomous, the first covariate variable *Education Level* is categorical with more than two response categories, and the second covariate variable *Income Level* is continuous. Through using these predictor variables with diverse levels of measurement, we can illustrate how to relate each to the dependent variable.

The subject matter involved in sample study one is purposefully less dire than it might be, as heavy subject matter tends to take attention away from the statistical methods. However, the selection of the proper statistical tests and procedures is largely based upon the level of measurement of the variables (categorical, continuous) and not the particular construct. Therefore, the analysis presented in Part 4 could apply to *any* study examining a continuous dependent variable.

4.2   Step 1: Study Map

Step 1 (Study Map) is the stage of the quantitative analysis where the study variables and hypothesized variable relationship(s) are listed in both text (**Section 4.2.1**) and diagram (**Section 4.2.2**) form.

4.2.1   The Study Map in Text

For sample study one, our study research question, hypothesis, dependent variable, independent variable, and covariate variables are:

Research Question: Are People who Live with Dogs Happier than People who Live with Cats?

Hypothesis: People who Live with Dogs Are Happier than People who Live with Cats

Dependent Variable: Happiness

Independent Variable: Do You Live with a Dog or a Cat?

Covariate Variables: Education Level, Income Level

4.2.2   The Study Map in a Diagram

Figure 4.1   Sample study one: Covariate, Independent, & Dependent Variables.

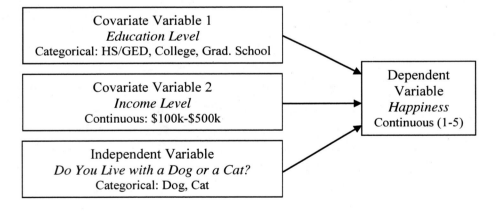

4.3   Step 2: Data Entry

Step 2 is Data Entry, which includes aspects of data coding, cleaning and recoding. Traditionally, data are collected by staff via interviews with study participants where hard copies (actual paper) of study surveys are completed. From this point, the completed study surveys are brought back to the research office where the data from the surveys are coded, entered, cleaned, and recoded if necessary. To best prepare the reader for the real world experience, in this section we will simulate this process using a short study survey based on our sample study.

4.3.1 Coding the Data

Figure 4.2 presents the study survey questionnaire for sample study one. Each item on this survey must be assigned a **code**, which will include a **variable name**, as well as **numbers for any assigned responses** to the item. The data provided by the study participants via the study survey can then be entered into the software database (e.g., the SPSS database) based upon these codes. In this section we will first observe the study survey (**Figure 4.2**), then the data dictionary (**Figure 4.3**), which presents each item on the study survey with assigned coding.

4.3.1.1 The data dictionary

The first step in entering the data is to create a data dictionary to reflect the codes assigned to variables. Each item on the hard copy of the study survey must be given a <u>variable name</u> that will represent that item in the software database. An example of a data dictionary is provided in **Figure 4.3**.

You may notice that the data dictionary is essentially the study survey with the coding reflecting the *variable names* assigned to each survey item, as well as a *number* to each potential response to that survey item superimposed. There are many ways to construct a data dictionary. However, many perceive that this is the best method because it provides one with a quick reference between the study survey and the corresponding items within the software database.

In the data dictionary, to make codings stand out from the regular survey items:

1) **All variable names are BOLDED, ALL IN CAPS in red** (However, in the SPSS database, variable names are typically not entered in all capital letters)

2) **All responses to items are** not bolded and not in caps in red.

Variable Names. There are no hard rules regarding the creation of variable names. It is best to assign a short name (approximately 8 characters) that effectively identifies that variable. For example, in the data dictionary within **Figure 4.3**, you will see that survey item #1 *Highest Education Level Completed* was assigned the variable name *educ* (which is typical). As mentioned, this type of name assignment must be repeated for each individual survey item.

Figure 4.2   The study survey questionnaire for sample study one

ID:

1. Highest Education Level Completed (circle one response):
   a. High School/GED
   b. College (4 Year Degree)
   c. Graduate School

2. Personal Income (Last Year): $_____(in thousands)

3. Have you ever had a drink of alcohol?
   a. If yes, then Indicate #Drinks Last Week ____Drinks
   b. If no, Continue to Item #4

4. Do you live with a (circle the appropriate response):
   a. Dog
   b. Cat

5. Tell us about your hobbies. Do you enjoy (circle all that apply):
   a. Skiing
   b. Surfing
   c. Reading

6. Circling the appropriate number:

In the last 7 days I felt:					
	Strongly Disagree	**Disagree**	**Neither Agree or Disagree**	**Agree**	**Strongly Agree**
Happy	1	2	3	4	5
Sad	1	2	3	4	5
Good	1	2	3	4	5
Strong	1	2	3	4	5
Glad	1	2	3	4	5

Remember in the case of an ALL THAT APPLY variable, each response category must be made into a separate variable with a response of Yes or No. For example, survey item #5 is an ALL THAT APPLY type survey item where study participants may select as many hobbies as they like (i.e., the item instructs *circle all that apply*). If survey item #5 was coded as a single variable where study participants could select one of three response categories (skiing, surfing, or reading), he or she could not make more than one selection. However, if each response category is made into a separate variable with a response category of Yes or No (as presented in the data dictionary in **Figure 4.3**), study participants may select any number (0, 1, 2, or 3) of hobbies listed.

Assigning Numbers to the Item Responses. After giving each survey item a variable name, you would then code each response to the item. For example, for survey item #1 *Highest Education Level Completed* (a.k.a., the variable *educ*) we might code the response category High School/GED as 1, College as 2, and Graduate School as 3. **Important**: When using certain measurement tools and instruments, survey items have specific numbers that must be assigned to each potential response. **It is important to keep this in mind and check for any pre-assigned coding**! The next paragraph will present an example of why this is essential.

The Importance of Adhering to Pre-Assigned Numeric Values and Cutoff Scores. Pre-assigned codes to item responses are often summed into a composite scale score with a designated **cutoff point**. A cutoff point is a value on a scale where scores above and below that value reflect some meaningful construct, such as a clinically significant level of depression. For example, a scale might contain 10 items measuring depression. These items may be combined into a composite score that includes a meaningful cutoff point that is based upon coding the item responses in a certain manner. For example, consider the scenario where:

- Each of the ten items receive a response of *Disagree*, *Neither Disagree or Agree*, and *Agree* that are coded with the values 0, 1, 2, respectively, as per the creators of the measurement tool. Based on these numbers, the potential range of summed scores from lowest to highest would be 0-20.

- Specifically, the lowest composite score of '0' would result from a study participant providing a response of *Disagree* (*Disagree* = 0) for all ten items, where 0 (for each response) multiplied by 10 (for the 10 items) is 0.

- The highest composite score of 20 would result from a study participant providing a response of *Agree* (*Agree* = 2) for all ten items, where 2 (for each response) multiplied by 10 (for the 10 items) is 20.

- Along this composite score continuum of 0-20, **a score of 10** is the cutoff score where scores of ten or greater (10-20) reflect a clinically significant level of depression.

- Consider what would occur if you did not observe the proper coding of 0, 1, and 2, but arbitrarily assigned the values 1, 2, and 3 for responses to each item.

- Using the coding 1, 2, and 3, for the 10 items, the lowest possible score would be 10 (as a score of 1 for each item multiplied by the 10 items is 10).

- Based upon the arbitrary 1, 2, 3, coding scheme, the lowest possible composite scale score (10) reflects a clinically significant level of depression as per the scale creators.

- Here a score of 10 is actually a score of 0 if the correct coding pattern was used, which reflects the lowest level of depression.

- However, due to the arbitrary coding, even those that indicate the lowest level of depression would receive a composite score of 10, which would be coded as above the cutoff point reflecting a clinically significant level of depression.

Yes or No Items. When coding items with a Yes or No response, it is generally preferable to code Yes as '1' and No as '0.' This is preferred because many times you will need to sum a series of Yes or No items where you would like to observe the number of Yes responses.

For example, suppose you wanted to sum the number of hobbies listed in survey item #5 (recall there were three hobbies listed).

- If Yes was coded as: '1' and No was coded as '2' (a. Skiing [1=Yes, 2=No] b. Surfing [1=Yes, 2=No] c. Reading [1=Yes, 2=No), then where a study participant responded No (=2) to all three hobbies and those three values were summed, a score of '6' (2 x 3 = 6) would be generated.

  Likewise, a score of '3' would be generated where a response of Yes (=1) was given for all three hobbies. It is often confusing to have a score of twice as much (6) where no hobbies were indicated relative to the score where all hobbies were indicated (3).

- However, when Yes is coded as '1' and No is coded as '0' (a. Skiing [1=Yes, 0=No] b. Surfing [1=Yes, 0=No] c. Reading [1=Yes, 0=No]), a score of '3' would be indicated where a response of Yes (=1) was given for all three hobbies, while a score of '0' would be assigned when a response of No (=0) was given for all three hobbies.

  Subsequently, it is often conceptually clearer to have the higher score reflect the higher number of occurrences (e.g., Yes responses), which occurs here when the No response is coded as '0' rather than '2.'

Figure 4.3   The study data dictionary

ID: [ID]

Please circle or enter the appropriate response to each item:

1. Highest Education Level Completed: [EDUC]
    a. High School/GED [1]
    b. College (4 Year Degree) [2]
    c. Graduate School [3]

2. Personal Income (Last Year): $[INCOME] (thousands) [Raw #]

3. Have you ever had a drink of alcohol? [DRINKALC]
    a. Yes [1] # Drinks Last Week [NUMDRINK] Drinks [Raw #]
    b. No [0]   Go to item #4

4. Do you live with a: [LIVEDOGCAT]
    a. Dog [1]
    b. Cat [2]

5. Please tell us about your hobbies. Do you enjoy (all that apply):
    a. Skiing [SKIING] [1=Yes, 0=No]
    b. Surfing [SURFING] [1=Yes, 0=No]
    c. Reading [READING] [1=Yes, 0=No]

6. Please circle the appropriate number:

In the last 7 days I felt:					
	Strongly Disagree	**Disagree**	**Neither Agree or Disagree**	**Agree**	**Strongly Agree**
Happy [FEELHAPPY]	1	2	3	4	5
Sad [FEELSAD]	1	2	3	4	5
Good [FEELGOOD]	1	2	3	4	5
Strong [FEELSTRONG]	1	2	3	4	5
Glad [FEELGLAD]	1	2	3	4	5

In summary, once variable names and numerical values have been assigned to survey items and the response categories (respectively), it is time to enter the variable names and numeric values into the software database. To do this, you must first create the SPSS database.

4.3.2   Entering the Data

Here we will present data entry as a two-step process. The first step is concerned with *Creating the Software Database* (**Section 4.3.2.1**), so data may be entered therein. The second step addresses *Entering Survey Data* (**Section 4.3.2.2**) where data from the hard copies of the study surveys completed by study participants are entered into the software database.

4.3.2.1   Creating the Software Database

To begin creating the SPSS database, open the SPSS program, go to **File, New, Data**. A new database will open. You will notice in the lower left hand corner of the screen that there are two tabs, **Data View** (used to enter data) and **Variable View** (used to create variables). The database will open in Variable View, which is the correct place to begin creating the database. From here, we will create the SPSS variables beginning with the most critical columns in the Variable View, which are the *Name, Type, Label, Values*, and *Missing* columns. These columns are critical because if they are not completed correctly, the statistical analysis will be compromised or not possible. Each of these columns is addressed in this section. In order to illustrate how to complete each of these columns, the first three survey items will be entered into Variable View of the database as variables. As per the study data dictionay (**Figure 4.3**) the first three variables are **ID** (Survey item: ID), **educ** (Survey item: 1. Highest Education Level Completed), and **income** (Survey item: 2. Peronal Income (Last Year)).

Column Labeled: *Name*

Within the **Variable View** portion of the SPSS database, under the column marked *Name* enter the variable names assigned to the survey item as per the study data dictionary (**Figure 4.3**). For example, **Figure 4.4** presents the variable names (**ID, EDUC**, and **INCOME**) assigned for the first three survey items.

Figure 4.4   The *Name* column in the SPSS **Variable View**

	Name
1	ID
2	educ
3	income

Column Labeled: *Type*

Under the column marked *Type* enter the type of variable, which is determined by the type of data being entered for that variable. If one clicks the right side of the cell marked *Type*, the dialogue box will open where you can select the variable type. You will see on the left hand side, one can select several types. All of the variables in our database are *Numeric* as the response categories are coded as numbers. Because all of the variables in our database are of the *Numeric* type, the column labeled *Type* will look similar to **Figure 4.5**. Variables can also be coded as several other types of responses when appropriate, as listed in the left hand column of the dialogue box. The most commonly used other categories are *String* (where text is entered) and *Date* (where calendar dates are entered). Please see the SPSS manual for a fuller description of these categories.

Figure 4.5   Adding the *Type* column in the SPSS **Variable View**

	Name	Type
1	ID	Numeric
2	educ	Numeric
3	income	Numeric

Column Labeled: *Label*

Under the *Label* column, enter a description of the variable (e.g., Level of *Happiness*). Many times the easiest label to enter is either the direct item from the study survey or a shorthand version of that item. For example, again regarding the first three survey items, you will see in **Figure 4.6**, under the column marked *Label* for the variable **ID**, the label simply states *Study Participant ID*. For the variable **educ**, the ascribed label is *Education Level* and for the variable **income** the label states *Income Level*.

Figure 4.6   Adding the *Label* column in the SPSS **Variable View**

	Name	Type	Label
1	ID	Numeric	Study Participant ID
2	educ	Numeric	Education Level
3	income	Numeric	Income Level

Column Labeled: *Values*

Under the *Values* column, we will enter the numerical coding for each response ascribed to a survey item. For example, let's enter the values ascribed to the study survey item #1 **Highest Education Level Completed**, which is coded as the variable *educ*. Recall as per the data dictionary (**Figure 4.3**) the variable *Education Level (educ)* has three potential responses, which are High School/GED (coded as '1'), College (coded as '2'), and Graduate School (coded as '3'). To enter the *values* in the SPSS database for this survey item, you would:

1) Go to the variable *Education Level* (variable name: *educ*) in the SPSS database in the **Variable View** portion.

2) Under the column labeled *Values*, click the right side of the cell for that variable to open the *Value Labels* dialogue box.

3) Enter a '1' in the **Value** cell, *High School/GED* in the **Label** cell, click **Add**.

4) Enter a '2' in the **Value** cell, *College* in **Label** cell, click **Add**.

5) Enter a '3' in the **Value** cell, *Graduate School* in **Label** cell, click **Add**.

6) Click **OK**.

The cell within the *Values* column will look similar to the intersecting cell for the variable *educ* as presented in **Figure 4.7**.

Figure 4.7   Adding the *Values* column in the SPSS **Variable View**

	Name	Type	Label	Values
1	ID	Nume	Study	None
2	educ	Nume	Educa	{1.00. High School/GED}…
3	incom	Nume	Inco	None

Column Labeled: *Missing*

Under the *Missing* column, we will enter the values that reflect the reasons for a missing data point (when a valid response has not been provided by the study participant for a survey item). The values we will use where a data point are missing should be extreme, so that the numbers will not be mistaken for valid responses. For example, if a study subject does not provide a value for his/her age (i.e., the study participant's age is missing), you need to enter a value in that empty cell. If you choose to enter '9' for missing values, it might be difficult to identify whether the respondent is actually nine years of age or if the age value was missing on the survey.

Therefore, it would be beneficial to code missing values as '999,' rather than only '9,' as '999' would not be a valid response for an item asking the age of study participants. Therefore, it is often useful to use extreme values for empty cells in the database. It is also important to realize that there are three types of missing values, which are missing, refused and not applicable, each of which requires a distinct numerical coding. For example, we might decide to code the missing values as:

1) The value 777 to indicate a Missing Value (777=Missing Value), where a response is simply not provided for an item.

2) The value 888 to indicate a Refused Item (888=Refused), where a study participant indicates he or she does not want to respond to an item.

3) The value 999 to indicate an item that is Not Applicable (999=Not Applicable), where an item does not apply to a study participant. For example, a response of *No* to survey item **3 Have you ever had a drink of alcohol?**, should be accompanied with code of *Not Applicable* (999) for next item asking **# Drinks Last Week [NUMDRINK]**.

These values are entered for each variable by:

1) Clicking on the right side of the cell in the *Missing* column for each variable.

2) Within the dialogue box that opens, enter the values 777, 888, and 999 in the three cells beneath the term **D̲iscrete Missing Values**.

3) Click **OK**.

Figure 4.8 presents the SPSS **Variable View** with the values for missing data points entered in the *Missing* column.

Figure 4.8   Adding the *Missing* column in the SPSS **Variable View**

	Name	Type	Label	Values	Missing
1	ID	Numeric	Study Participant ID	None	777, 888, 999
2	educ	Numeric	Education Level	{1.00. Hi	777, 888, 999
3	income	Numeric	Income Level	None	777, 888, 999

The Remaining Columns

Within the SPSS **Variable View**, after the primary columns are completed, a number of other columns remain, which are the *Width, Decimals, Columns, Align, Measure,* and *Role* columns. Although these are not described here as primary columns, these columns can be important and useful. Feel free to refer to the SPSS manual for a full presentation on these columns.

4.3.2.2 Entering survey data

Figure 4.9 presents a completed study survey that we will now enter into the SPSS database.

Figure 4.9   The completed study survey used for data entry

ID: 1

1. Highest Education Level Completed:
~~High School/GED~~ (circled)
~~College (4 Year Degree)~~
Graduate School

2. Personal Income (Last Year): $___125,000___(in thousands)

3. Have you ever had a drink of alcohol?
    a. If "Yes" then Indicate # Drinks Last Week ___4___Drinks
    b. If "No" to item #3: Please go to item #4

4. Do you live with a:
    a. Dog (circled)
    b. Cat

5. Tell us about your hobbies. Do you enjoy (circle all that apply):
    a. Skiing (circled)
    b. Surfing
    c. Reading

6. Circle the appropriate number:

In the last 7 days I felt:					
	Strongly Disagree	**Disagree**	**Neither Agree or Disagree**	**Agree**	**Strongly Agree**
Happy	1	2	3	4	(5)
Sad	(1)	2	3	4	5
Good	1	2	3	(4)	5
Strong	1	2	3	4	(5)
Glad	1	2	3	(4)	5

After the *Name, Type, Label, Values,* and *Missing* columns are structured (as well as any other columns desired) the data from the hard copy of the survey can be entered into the **Data View** portion of the database. To begin, use the tab in the lower left hand corner of the SPSS screen to switch into the **Data View** screen. Once in the **Data View** portion of the database, you will notice that the name of each study variable is at the top of each column. Here is where you can begin to enter the data from the study survey hard copies. To do this in SPSS:

1) Look at the completed study survey presented in **Figure 4.9**.

2) Note the value entered for the first coded survey item, which is the value '1' provided for the survey item *ID (Coded as: ID)*. Note this survey item corresponds to the SPSS variable *Study Participant ID (Variable name: ID)*.

3) Subsequently, in the **Data View** portion of the SPSS database, enter a value of '1' in the column labeled *ID* as presented in **Figure 4.10**.

4) Next, go down the completed survey and enter the values from the survey into the corresponding SPSS variable. For example, note on the survey the next item is the item *1. Highest Level of Education Completed* (SPSS variable name *educ*), which has the response *High School/GED* circled.

5) Subsequently, you would enter a '1' (for the variable *educ*, 1=High School/GED) in the *Data View* screen under the variable name *educ* as presented in **Figure 4.10**.

6) Follow this procedure until all the data from the survey hard copies are entered for each study participant.

Figure 4.10   Entering the raw data into the **Data View** portion of the SPSS database

	ID	educ	income	drinkalc	numdrink
1	1.00	1.00	125000	1	4
2					
3					

Once all the data are entered from the completed survey hard copies, they will be ready for cleaning and then recoding as described in the next two sections.

4.3.3 Cleaning the data

Cleaning the data refers to the process of reviewing the data in your database to detect, correct, or remove incorrect, inconsistent, or inaccurate values, which are commonly referred to as *dirty data*. These dirty data have several causes such as data entry errors, corruption in transmission or storage, or invalid responses given by study participants. Data cleaning is tremendously important, as this process is our means of being certain that the data present in the software database actually reflect the responses given by study participants via data collection. In this section we will discuss three of the essential methods used to clean data. First, we will discuss *Referencing Survey Hard Copies* (**Section 4.3.3.1**) to the data that have been entered into the software database, which is a great method toward detecting errors in data entry. Then we will discuss *Examining the Variables* (**Section 4.3.3.2**) and *Violations in Logic* (**Section 4.3.3.3**), which are both useful on many levels, including the detection of inappropriate responses provided by study participants.

4.3.3.1 Referencing survey hard copies

The most comprehensive way to check if the data from the hard copies of the study surveys are entered correctly into the SPSS database is to check each response from the paper surveys against what is entered for each response in the SPSS database. For example, if we want to be sure that our completed study survey #1 was entered into the SPSS database correctly for the item *Highest Education Level Completed* we would:

1) Retrieve the hard copy of study survey #1.

2) Open the SPSS database in Data View and find '1' in the ID variable (which corresponds with the study survey #1 hard copy).

3) See that the response for the survey item #1 (Highest Education Level Completed) is High School/GED.

4) The data dictionary tells us that the response category for High School/GED is '1' and that the survey item corresponds to the variable *educ* in the SPSS database.

5) Then look at the SPSS database to check that for ID #1, under the variable *educ*, the number '1' (=High School/GED) is entered in that cell.

6) Repeat this check between the responses on the hard copy of the study survey and the values in the SPSS database for the rest of the items.

At times it may not be practical to check all of the hard copies of the study surveys against all the data entered in the SPSS database. In such instances, you might take a random sample of the completed surveys (e.g., 10% where you would select every 10th study survey for review) to perform these checks.

4.3.3.2 Examining the variables

The next step in cleaning the data is to examine the variables. There are several commonly used methods applied to examine the variables at this step including boxplot and histogram graphs, which we will review later in the text. However, perhaps the most simple and straightforward way to examine the variables is to run a **frequency procedure** for each variable. The frequency procedure displays the values that have been entered for each item in the software database. Subsequently, the frequency procedure is useful in cleaning the data as the results will indicate if an **inappropriate or invalid response** has been entered in the database for a survey item/variable in the database. For example, let's consider the first item in the 5-item Happiness scale, which is the variable item: *In the last 7 days I felt: Happy* (SPSS variable name = *feelhappy*). We know:

1) The potential range of scores for this item is 1 (1=Strongly Disagree) to 5 (5=Strongly Agree).

2) Therefore, if we conduct a frequency procedure for this item, we should only see values entered for this item between 1 and 5.

3) Any numerical values entered for this variable outside this interval of 1 to 5 (except for the values indicating empty cells, which are 777, 888, 999), would likely be data values that were not entered into the SPSS database correctly or were invalid responses reported by study participants.

4) To conduct the frequencies procedure to examine the variable item *In the last 7 days I felt: Happy* (SPSS variable name = *feelhappy*) in SPSS go to: **Analyze→ Descriptive Statistics→ Frequencies**.

5) In the dialogue box that opens (see **Figure 4.25**), move the variable named *feelhappy* (Survey item = *Last 7 Days: Feel Happy*) from the column on the left into the column on the right.

6) Then click **OK** or **Paste**.

The statistical output will look similar to **SPSS Output 4.1**. Notice, the values provided for the variable *feelhappy* are listed within the column on the far left. Here you will see either raw values, or the values that have been entered for each response for the item. For example, within the output, within this column, instead of the raw numbers 1, 2 3, 4, and 5, that have been entered in the SPSS database, we see the corresponding **Values** to these numbers, which are *Strongly Disagree* (=1), *Disagree* (=2), *Neither Agree or Disagree* (=3), *Agree* (=4), and *Strongly Agree* (=5).

A number entered as a response for this variable that is not a number assigned a **Value**, will appear as a raw number, which is often an inappropriately entered value for that variable. For example, in the **SPSS Output 4.1**, we see the value 33.00 in the left hand column. This number is not assigned a **Value** and is outside the potential range of scores (1 to 5), as well as the values reflecting an empty cell (777, 888, and 999). Thus, it appears that the value 33.00 is an incorrectly entered data point that needs to be corrected. It is likely that the correct value was a 3.00, but the person performing data entry pressed 3 twice by mistake.

SPSS Output 4.1

Last 7 Days: Feel Happy

		Frequency	Percent	Valid Percent	Cumulative Percent
Valid	Strongly Disagree	11	11.0	11.0	11.0
	Disagree	38	38.0	38.0	49.0
	Neither Disagree or Agree	23	23.0	23.0	72.0
	Agree	19	19.0	19.0	91.0
	Strongly Agree	8	8.0	8.0	99.0
	33.00	1	1.0	1.0	100.0
	Total	100	100.0	100.0	

33.00 is outside the 1 to 5 range, so it is not a valid response

The first step in correcting a data point, is to check if the correct value was entered from the survey hardcopy by referring that document. This can be done by linking the value entered for the **ID** variable in the SPSS database to the matching number in the **ID** field on the study survey hard copy. This first step is presented in **Figure 4.11**, where we have referenced the **Data View** portion of the SPSS database to find the **ID** number linked to the response entry of 33.00 (for the variable *feelhappy*). We see **ID** number '3' is the case associated with the value 33.00.

Figure 4.11   Locate the invalid response to obtain the participant ID number

ID	feelhappy
1.00	3.00
2.00	4.00
3.00	33.00

Find the Study Participant with the response value of 33.00, which is **ID 3**

Find the invalid response (33.00) in *feelhappy*

The next step would be to reference the study survey hardcopy to see if the value 33.00 was a typo or if that was the value provided by the study participant. **Figure 4.12** presents the study survey hardcopy, for **ID** number '**3**' that reflects a response of '3' was the actual response given by the study participant. Therefore, it would be reasonable to assume that the dirty data point of 33.00 in the SPSS database for the variable *feelhappy* (for **ID** 3) was a typo derivative of the process of data entry. Subsequently, it would be appropriate to return to the **Data View** portion of your SPSS database and replace the value 33.00 with a 3.00, for study participant **ID** 3, within the variable *feelhappy* as presented in **Figure 4.13** below.

Figure 4.12 Referencing the study survey hardcopy with the SPSSS variable

ID 3

Happy (Variable: *feelhappy*) has a response of '3' on the survey hard copy

Study Survey ID: 3

In the last 7 days I felt:					
	Strongly Disagree	**Disagree**	**Neither Agree or Disagree**	**Agree**	**Strongly Agree**
Happy	1	2	(3)	4	5
Sad	(1)	2	3	4	5

Figure 4.13 Correcting the dirty data (mistyped) point

ID	**feelhappy**
1.00	3.00
2.00	4.00
3.00	3.00

For **ID** 3, 33.00 is replaced with the accurate 3.00

Next, the frequencies procedure should be repeated for each variable within the SPSS database to inspect the data problems.

4.3.3.3 Violations in logic

Another key safeguard that should be taken when cleaning the data is to look for violations in logic. A violation in logic occurs when a response given to one item by a study participant conflicts with a response he or she gave to another survey item. An example of this is where a skip pattern has been violated. In our sample study survey, Item #3 is a skip pattern item where study participants are asked "**Have you ever had a drink of alcohol**?" If the study participant answers *Yes*, he or she is then asked to **Indicate the # Drinks Last Week _____Drinks**. Many times when you have a skip pattern like this, a study participant might respond "No" to item #3 (No, as in never had a drink of alcohol), but then will enter a number for the question **# Drinks in the last week**. In the SPSS database, these variables will look similar to **Figure 4.14**, where the highlighted responses indicate a study participant reported *No* to the item **Have you ever had a drink of alcohol**?, but '6' to the item **# Drinks Last Week**.

Figure 4.14   A violation in logic: Conflicting responses in a skip pattern

drinkalc	numdrink
No	999
Yes	4.00
No	6.00

Reports never had a drink of alcohol ⟶ No

But also reports having ← 6 drinks last week!

In the event that this occurs, go back to the survey hard copy and check if the actual response given is reflected in the database, and that no data entry error occurred. If the response conflict reflected within **Figure 4.14** are the values provided by the study participant, the data are most often treated as *dirty data*. Dirty data in this instance would typically be <u>unusable</u>. For example, when the answers conflict as they do above where the study participant reported that he/she never had a drink of alcohol, but had 6 drinks last week, the data are not usable because we do not know which response is correct.

Perhaps the study participant meant to circle "Yes" for item #3 and report 6 drinks. Or perhaps the study participant meant to circle "No" for item #3 and the '6' entered for the variable *numdrink* was a mistake. Since we cannot tell, we cannot use the data. Therefore, the responses given should be coded as *missing* by entering a *777* in the cells replacing the dirty data. *The concern for data fidelity should be paramount and we should never assume that we know what a study respondent was trying to say, only a consideration of what was indicated*.

The Importance of Identifying Violations in Logic in the Data Prior to Data Analysis. It is critical to identify and recode dirty data due to violations in logic before beginning data analysis. One of the main reasons is that study participant responses might look valid and be included in the data analysis, leading to erroneous conclusions. Suppose after entering the data we failed to compare responses to item #3 to the subsequent item that asks # Drinks Last Week. Instead we perform an analysis that examines factors that predict item #3 *Have you ever had a drink of alcohol? (Yes or No)*. If we only considered the Yes or No response to item #3 without checking for conflicts with the next item, then the responses to item #3 look valid even if there is a conflict. Subsequently, we might conduct an entire analysis that yields interesting findings. We may even write a manuscript on these findings or present them at a conference.

Then as we go back and examine the data, we might come around to comparing item #3 to the next item (#Drinks Last Week) and realize there are several discrepancies between the two items and several responses must be recoded as missing since they are unusable dirty data. When we re-run the analysis with the unusable responses removed, we may find our original findings have changed and are no longer statistically significant. As a result, the manuscript written based upon these findings might be discarded since the findings may no longer be valid. You might have to also inform any pertinent parties involved that the initial findings were based upon a methodological error. This is only one of many reasons describing why it is imperative to clean data prior to data analysis.

4.3.3.4 Recoding Variables

Recoding variables generally refers to changing a variable in some fundamental way to suit the purpose of the study. For example, if letters (e.g., A, B, C) were given as responses to an item, the text needs to be recoded into numbers to facilitate a statistical analysis. In this section, we will mention two of the most common techniques used to recode variables, which are ***Reverse Coding Variables*** and ***Dummy-Coding Variables***.

Reverse Coding Variables. Reverse coding variables is commonly needed at some point within most studies. For example, as we reviewed earlier, many study measurement tools use a composite score that are composed of several individual items that need to be combined to measure a construct. Many times these multi-item scales contain items that are positively phrased (where a higher item score indicates a greater presence of the construct measured by the scale) and negatively phrased (where a higher item score indicates a lesser presence of the construct measured by the scale).

For example, **Figure 4.15** presents the directionality (i.e., if an item response indicates a greater or lesser degree of *Happiness*) of the five questions that combine to form the composite scale score measuring the construct *Happiness*. Notice within this scale that a responses of '5' (Strongly Agree) indicates a greater presence of the construct *Happiness* for the items *Happy*, *Good*, *Strong*, and *Glad*. The response reflecting the highest level of *Happiness* for each item is indicated by the yellow smiley face. However, a response of '5' for the item *Sad* reflects the lowest level of the construct *Happiness*, as a response of '5' reflects the highest rating of feeling *sad* (indicated by the red face frowning). In fact a response of '1' for the item *sad* reflects the highest presence of *Happiness*, as a score of '1' indicates the lowest rating of feeling *sad*.

Before the five questions that compose the 5-item composite scale measuring *Happiness* can be combined, all five questions must have the same pattern, where the high score for each item reflects the highest presence of *Happiness*. Therefore, the item *Sad* <u>needs to be reverse coded</u> where a score of '5' indicates a greater presence of the construct *Happiness* to match the pattern of the other four items. In other words, we must move the yellow smiley face to the other end of the item *sad*. The necessary recode of values for the item *Sad* is presented in **Figure 4.16**.

Figure 4.15   The directionality of the five items on the *Happiness* scale

In the last 7 days I felt:					
	Strongly Disagree	**Disagree**	**Neither Agree or Disagree**	**Agree**	**Strongly Agree**
Happy	1	2	3	4	5=  🙂
Sad	1  = 🙂	2	3	4	5=  ⚫
Good	1	2	3	4	5=  🙂
Strong	1	2	3	4	5=  🙂
Glad	1	2	3	4	5=  🙂

Figure 4.16   Needed recode before the *Happiness* composite score can be computed

Original Value for the item *Sad*	Changed to Reverse Coded Value
1 (Strongly Disagree)	⟶ 5
2 (Disagree)	⟶ 4
3 (Neither Agree or Strongly Disagree)	⟶ 3
4 (Agree)	⟶ 2
5 (Strongly Agree)	⟶ 1

To recode a variable, such as *feelsad*, in SPSS go to:

1) **Transform→ Recode into Different Variables**.

2) Move the variable to be recoded (*feelsad*) from the column on the left into the column on the right.

3) Under **Output Variable**, enter the recoded variable name into the cell **Name**. For example, here we have entered the name *feelsadREV*.

4) Enter a description in the **Label**, such as *Item Feel Sad Reversed*.

5) Click **Change** to move the new variable name into the center column (see **Figure 4.17**).

Figure 4.17   Reverse coding the item *feelsad* into *feelsadrev*

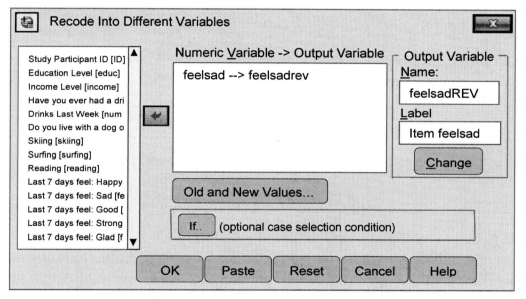

6) Next, click **Old and New Values**.

7) In the **Old Value** column, for **Value** enter the number '1,' then in the **New Value** column for **Value** enter a '5' and click **Add**.

8) In the **Old Value** column, for **Value** enter the number '2,' then in the **New Value** column, for **Value** enter a '4' and click **Add**.

9) In the **Old Value** column, for **Value** enter the number '3,' then in the **New Value** column, for **Value** enter a '3' and click **Add**.

10) In the **Old Value** column, for **Value** enter the number '4,' then in the **New Value** column, for **Value** enter a '2' and click **Add**.

11) In the **Old Value** column, for **Value** enter the number '5,' then in the **New Value** column, for **Value** enter a '1' and click **Add**.

12) The dialogue box should look similar to the **Figure 4.18**.

13) Click **Continue**.

14) Click **OK** or **Paste**.

Figure 4.18 Old and New Values: Recoding *feelsad* into *feelsadrev*

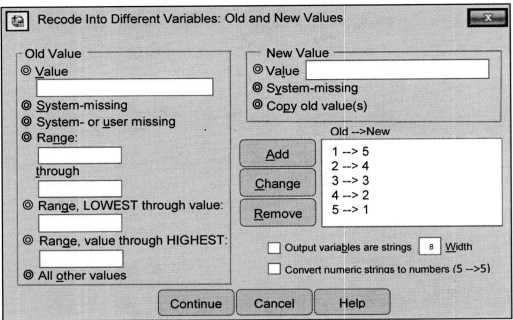

If you go to the end of the variable list in either the **Data View** or **Variable View** portion of the SPSS database, you should find the newly created reverse coded variable *feelsadREV*. In the Variable View portion of SPSS, feel free to enter a label if needed, as well as the values adjusted for the recoded values (e.g., 1=Strongly Agree, 2=Agree, 3=Neither Agree or Disagree, 4=Disagree, 5=Strongly Disagree).

See **Figure 4.19** for a conceptualization of how the item *sad* has been recoded to match the pattern of the other four items forming the 5-item *Happiness* composite scale. Within the figure, you may notice that the recode has moved the smiley face from one end of the item scale to the other, where a response of '5' now indicates a greater presence of the construct *Happiness*.

In other words, the recoded item *Sad* now matches the pattern of the other four scale items (*Happy, Good, Strong, Glad*) where a response of '5' now indicates the greatest presence of the construct *Happiness*.

Figure 4.19   The directionality of the item *sad* after *reverse coding*

In the last 7 days I felt:					
	Strongly Disagree	**Disagree**	**Neither Agree or Disagree**	**Agree**	**Strongly Agree**
Happy	1	2	3	4	5= ☺
Sad	1=5 ●	2=4	3=3	4=2	5=1 ☺
Good	1	2	3	4	5= ☺
Strong	1	2	3	4	5= ☺
Glad	1	2	3	4	5= ☺

Whenever a variable is recoded it is essential to check that the coding was successful. One means of checking if a reverse coding is successful is to run the frequency procedure on the original and recoded version of the variable and observe if the frequency of scores is reversed for the new reverse coded variable. To do this, in SPSS go to:

1) **Analyze→ Descriptive Statistics→ Frequencies**.

2) In the dialogue box that opens (see **Figure 4.25**), move the original (*feelsad*) and reverse coded (*feelsadREV*) version of the variables from the column on the left into the column on the right.

3) Click **OK** or **Paste**.

The two frequency tables presented in **SPSS Output 4.2** and **4.3** will be produced. **SPSS Output 4.2** presents the first frequency table for the original version of the variable *feelsad*. Within this output, we see there is a frequency of:

- 8 in the category *Strongly Disagree*

- 25 in the category *Disagree*

- 49 in the category *Neither Disagree or Agree*

- 12 in the category *Agree*

- 6 in the category *Strongly Agree*

SPSS Output 4.3 presents the second frequency table for the reverse coded version of the variable *feelsadREV*. Within this figure, we see a frequency distribution that is exactly the opposite of the original variable with:

- 6 in the category *Strongly Disagree*

- 12 in the category *Disagree*

- 49 in the category *Neither Disagree or Agree*

- 25 in the category *Agree*

- 8 in the category *Strongly Agree*

You will notice that the distribution within the original variable from the top to bottom (8, 25, 49, 12, 6) is inverted in the second reverse coded version of the variable (6, 12, 49, 25, 8).

SPSS Output 4.2

Last 7 Days: Feel Sad

		Frequency	Percent	Valid Percent	Cumulative Percent
Valid	Strongly Disagree	8	8.0	8.0	8.0
	Disagree	25	25.0	25.0	33.0
	Neither Disagree or Agree	49	49.0	49.0	82.0
	Agree	12	12.0	12.0	94.0
	Strongly Agree	6	6.0	6.0	100.0
	Total	100	100	100.0	

SPSS Output 4.3

Last 7 Days: Feel Sad (REVERSED)

		Frequency	Percent	Valid Percent	Cumulative Percent
Valid	Strongly Disagree	6	6.0	6.0	6.0
	Disagree	12	12.0	12.0	18.0
	Neither Disagree or Agree	49	49.0	49.0	67.0
	Agree	25	25.0	25.0	92.0
	Strongly Agree	8	8.0	8.0	100.0
	Total	100	100	100.0	

Dummy-Coding Variables. Dummy-coding is another common recode used in quantitative studies. Dummy-coding commonly refers to transforming variables into a dichotomous variable (a categorical variable with two response categories), such as Yes or No. There are many situations that would require dummy-coding variables. For example, when conducting the multivariate procedure known as regression, which we will do in **Step 6: Multivariate Analysis** of our sample studies, a categorical predictor variable with more than two response categories must be dummy-coded. For example, the categorical covariate variable *Education Level* has three response categories of: High School/GED, College, and Graduate School. Therefore, to use this variable in a regression model the response categories must be dummy-coded.

Since this dummy-coding process is linked with conducting regression, the dummy-coding process is presented in detail in **Step: 6 Multivariate Analysis**. Specifically, in this section the dummy-coding procedure is applied to the categorical covariate predictor variable *Education Level* which has more than two (three) response categories. Thus, to reference the instructional materials on dummy-coding please refer to section **4.7.1 Dummy-Coding Variables**.

4.4   Step 3: Checks of Data Integrity

Step 3: Checks of Data Integrity is concerned with one central question: **Are the Data <u>Appropriate</u> for Data Analysis?** This is perhaps the most important question of any study, even more so than the research question(s) and hypothesis(es). Why? If the data are not appropriate for data analysis, the research question(s) and hypothesis(es) are likely to be answered incorrectly!

I had an old football coach who used to say: *You can't start with chicken feathers and get chicken salad.* Thus, following the coach's culinary logic, even if you follow the world's best recipe for a chicken salad sandwich perfectly, the sandwich won't be of much use if it is made of chicken feathers. In the same way, even if you perform a data analysis procedure perfectly, if you are using data that are not appropriate for analysis, the results of that analysis are about as useful as a chicken salad sandwich made of chicken feathers (i.e., dubious at best). Obviously, it is essential to perform these checks of data integrity well in order to produce a fine product.

Additionally, it would be ideal to report the results of the checks of data integrity in detail within your research report (e.g., peer-reviewed manuscript), so readers know you are not feeding them chicken feathers. Keep in mind, a thorough check of data integrity is often the hallmark of a sophisticated and professional data analysis study. Therefore, a detailed description of how checks of data integrity were performed builds confidence in a study.

This chapter details each check of data integrity, which are statistical power (**Section 4.4.1**), test assumptions (**Section 4.4.2**), missing data (**Section 4.4.3**), and measurement tools (**Section 4.4.4**).

4.4.1 **Statistical power**

If I told you that I was going to conduct our sample study examining whether people who live with dogs or cats have a significantly different level of *Happiness*, you might ask: Just how many people are you going to include in this study?

In response, I might tell you: *Overall, I have 10 people, 5 people who live with dogs and 5 people who live with cats.* You might then respond that this seems like too few people in the study. Subsequently, I may suggest: *What if I gather data from 10,000 people, 5,000 people who live with dogs and 5,000 people who live with cats?* You might note that this seems like more people than necessary for the study. So, now we are both stuck wondering how many people we need for this study. Obviously, the right number is somewhere between 10 and 10,000. But how can we come up with a specific number?

Enter the power analysis! The power analysis is the primary tool in statistics used to determine how many study participants you need for your study. How does the power analysis come up with this number? Put simply, the power analysis indicates the number of study participants needed to generate enough statistical power to examine the relationships between study variables, such as the independent and dependent variables.

For example, if we would like to relate an independent variable such as *Do You Live with a Dog or a Cat?* with a dependent variable such as *Happiness*, we obviously need a certain amount of study participants. The certain amount of study participants we need is determined by the amount of statistical power needed to relate those variables to one another via a statistical test. The power analysis procedure only asks us for a few settings (e.g., power, alpha, effect size) before the specific number of study participants needed for a data analysis study is given.

4.1.1.1 Power Analysis

A power analysis is perhaps best conceptualized as a dish composed of four ingredients, which are settings for power, alpha, effect size, and sample size (for a fuller discussion see Cohen, 1988, 1992). It is important to realize that these issues surrounding statistical power estimates and the implications on statistical findings are wider than our discussion here (Kraemer & Thiemann, 1987; Wolins, 1982), but in brief, we can describe our power analysis as a function of setting:

Power: You must first set the value of power, which is defined as the ability to detect a significant effect if a real difference exists. Typically, .80 (80%) is considered an acceptable value for power. Subsequently, we will use power set at .80 for our sample study calculation. When higher values for power are used (e.g., .90), there is a greater ability to detect an effect, but the sample size required will increase.

Alpha: Alpha here, just as in inferential statistical tests, refers to the level of probability we are willing to accept that the findings produced are by chance. Furthermore, just as in our inferential statistical tests, alpha is typically set at .05. This is the value we will use in the power analysis for our sample study. Setting alpha at .05 ($p<.05$) suggests with the given sample size, that if there is not really an effect between the variables, the specified effect size would be seen in only 5% of studies.

Effect Size: The term 'effect size' refers to the magnitude of the effect between variables. Typically, an effect size will be categorized as being small, medium, or large. The smaller the specified 'effect size' in a study, the larger the sample size will be needed to detect that effect. There are several factors involved with selecting an effect size for a power analysis. However, among studies in general, if there is not a compelling reason to select a small or large effect size, the effect size is designated as medium. Subsequently, the power analysis for sample study one will use a medium effect size.

Sample Size: In a power analysis, the sample size required for the study is most often calculated based upon the other three settings (the power, alpha, and effect size).

Power Analysis Using the G*Power Software Program. There are many power analysis software programs available on the web, many of which are reasonably priced and/or give free limited time trial usage. There are also programs such as *G*Power* that are available for use at no charge. You can access the G*Power statistical software program at: http://www.psycho.uni-duesseldorf.de/abteilungen/aap/gpower3/.

The current power analysis utilizes the G*Power statistical software program. **Figure 4.20** presents the dialogue box based upon the current version of the G*Power statistical software program. Within the dialogue box, in the upper left hand corner you can select the statistical test that the power analysis will be based upon. Within this option, the test of Multiple Linear Regression has been chosen, as this is the multivariate test used in sample study one. Next, within **Figure 4.20**, we will enter the values for the three specified settings of power, alpha, and effect size, before we activate the *Calculate* button to produce an estimate of the needed sample size for the analysis. We will now discuss entering the values for each of these settings within the dialogue box.

Effect Size: The first line reads *Effect size f² ,* which is where we will specify the effect size. The line at the bottom of the dialogue box reads *Effect Size Conventions.* Here the effect size conventions, in this instance for a Multiple Linear Regression Model, are presented. Specifically, a medium effect size is specified as 0.15. Subsequently, in the cell to the right of the line *Effect size f²* we have entered 0.15.

Alpha: The second line reads *Alpha,* which is where we will specify our specified level of alpha, which is .05. Thus, in the cell to the right of the term *Alpha* reads 0.05.

Power: The third line reads *Power,* which is where we will specify our level of power, which is .80. Thus, the cell to the right of the term *Power* reads 0.80.

Predictors: The fourth line that reads *Predictors.* In a power analysis for a Multiple Linear Regression Model, there is often a need to *estimate* of the number of predictors that will be included in the final model. In our power analysis, we have estimated four.

Sample Size: Once we have specified our settings of effect size, alpha, and power, as well as our number of predictors, we are ready to click the *Calculate* button in the upper left hand corner and observe the estimated required sample size for our study.

Figure 4.20   A power analysis for a linear (multiple) regression model

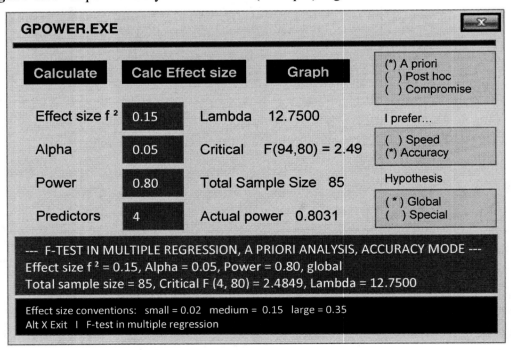

Figure 4.20 displays the results of this calculation, within the area labeled *F-TEST IN MULTIPLE REGRESSION, A PRIORI ANALYSIS, ACCURACY MODE* in the lower portion of the dialogue box. The final line of this area reads *Total sample size* = 85.

This is our sample size estimate. Specifically, the software indicates that a sample size of 85 study participants is required to detect a medium size effect in a multiple linear regression model including four predictors, with power set at .80 and alpha set at .05.

Why Only One Power Analysis? You might notice that we did one power analysis for sample study one, which was based on a Multiple Linear Regression Model. However, you will see later in this study we also perform other statistical procedures at the bivariate level such as a t-test, ANOVA, and Correlation, which are all inferential tests that also require a sufficient level of statistical power. In fact the G*Power tool also has settings for all three procedures, so you might rightfully ask why did we do a power analysis for our final multiple linear regression analysis, but not for these other tests? The answer is that it is quite typical to include a power analysis in a grant, manuscript, research report, and so on, based solely on the final multivariate model.

Summary. The power analysis presented in this section is about as straightforward as this procedure can be. Many types of power analysis procedures require estimates of mean and standard deviation values, as well as attrition (the proportion of study participants who exit a study prior to completion), which makes the power analysis a bit more involved. However, our presentation is meant to be more of a foundational grounding in the procedure as the issue of statistical power and the power analysis is an important check of data integrity.

4.4.2 Test assumptions

All quantitative models have certain assumptions that must be reflected within the data before a statistical procedure can be applied. Violations of these assumptions can lead to statistical findings that misrepresent the study data (Pedhazur, 1997, p. 33). There are five principal assumptions that must be met before the multiple linear regression procedure can be used in analysis, which is the final multivariate model to be used in **Step 6: Multivariate Analysis**. These five assumptions, along with several other smaller assumptions are presented in the six subsections within **Section 4.4.2**.

Specifically, this section presents six subsections that describe essential test assumptions that must be addressed when performing multiple linear regression, including a normal distribution (**Section 4.4.2.1**), multicollinearity (**Section 4.4.2.2**), homoscedasticity (**Section 4.4.2.3**), linearity (**Section 4.4.2.4**), no undue influence of outlier scores (**Section 4.4.2.5**), and other test assumptions (**Section 4.4.2.6**). In each section we will discuss what the assumption is, how to test to see if the assumption has been met, and what to do if testing indicates that the assumption has not been met.

4.4.2.1 Normal Distribution

Parametric tests, such as linear regression, assume that a continuous variable has a distribution of scores that approximates a **normal distribution**. Once you have affirmed that the distribution(s) of your continuous variable(s) is approximately normal, you can move on to the next test assumption. However, this assessment has the potential to be rather lengthy. Specifically, if continuous scores are normally distributed or easily adjusted, this assessment will be rather fast. However, at times, a distribution may be significantly non-normal and may need considerable attention. We will review several scenarios as a means of preparing for addressing the question of normality. Specifically, in this section we will review this information in the following four parts:

Part A: What is a Normal Distribution of Scores?

Constructing Your Own Normal Distribution of Happiness Scores

Part B: What Are the Threats to a Normal Distribution?

Threat 1: Skewness

Threat 2: Kurtosis

Threat 3: Outlier Scores

Part C: How Do You Assess if Scores are Approximately Normally Distributed?

Descriptively: Examine the Score Distribution Using Frequency Counts and Charts

Statistically (1): The Kolmogorov–Smirnov and Shapiro–Wilk Statistical Tests

Statistically (2): The Ratio of Skewness and Kurtosis to the Standard Error of Each

Part D: How Do You Address Problems Related to a Non-Normal Distribution?

Option 1: Assessing the Impact of Outlier Scores (Steps 1-4)

Option 2: Data Transformations

Part A: What is a Normal Distribution of Scores?

A normal distribution of scores is a theoretical distribution of values that makes a bell shaped symmetrical curve (Nicol, 2010). Please see **Figure 4.21** for an illustration. Within this figure, the '0' in the center bottom represents the mean score. The numbers to the left and right of the '0' indicate units of standard deviation, which describe the percentage of cases that will vary from the mean score in 1, 2, 3, and 4 units of standard deviation if a distribution is perfectly normal. For example, we see that 68% of scores will fall within one standard deviation of the mean score. This is explained in greater detail later in this section. For a review on the definitions of the mean and standard deviation values, please refer to **Section 4.5 Step 4: Univariate Analysis**.

Figure 4.21   A Normal Distribution of Scores

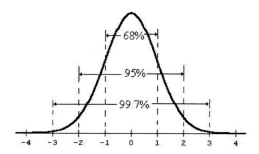

However, rather than defining a normal distribution, it may be more productive to describe how you already know what a normal distribution of scores is. Suppose I told you that I had administered a scale measuring *Happiness* to 100 people encountered at random on a street in New York City. Before I even described the collection of 100 *Happiness* scores, you would likely have three intuitive assumptions. Specifically, you might anticipate there will be a group of people that are significantly unhappy (i.e., a group with particularly low *Happiness* scores), which we will call **Intuitive Assumption A**. There will be a group of people that are significantly happy (i.e., a group with particularly high *Happiness* scores), which we will call **Intuitive Assumption B**. However, most people will be in the middle where they are not especially happy or unhappy (i.e., most people will evidence somewhat average *Happiness* scores), which we will call **Intuitive Assumption C**.

These assumptions reflect the idea of a normal distribution of scores. The 68% of scores are "clustered" around the mean score and are generally considered to be similar to the mean score. The remaining 32% of scores are outside one standard deviation of the mean score. These scores are generally considered to be significantly different from the mean value score. Of these 32% of scores, 16% are above (to the right) and 16% are below (to the left) one standard deviation of the mean value score. Regarding *Happiness* scores, the three groups of 16% (significantly low scores), 68% (the large group of similar scores), and 16% (significantly high scores) reflect intuitive assumptions A, B, and C.

Constructing Your Own Normal Distribution of Happiness Scores. In fact, if you know the *mean* score and *standard deviation* (SD) of a sample, you can use this information to construct a normal distribution to identify which scores within a continuum of scores (such as the *Happiness* scores measured from **1-5**) reflect these three groups. For example, the mean score for *Happiness* is **2.73** and the standard deviation is **.76**. To construct a normal distribution of *Happiness* scores that identifies these three groups, we would:

1) Identify one SD above the mean score by adding one unit of SD (.76) to the mean score (2.73), which is 3.49.

2) Identify one SD below the mean score by subtracting one unit of SD (.76) from the mean score (2.73), which is 1.97.

3) Thus, regarding *Happiness* scores, 68% of scores should be between 1.97 and 3.49, which is the interval within one SD of the mean score. These scores reflect **Intuitive Assumption C**, which is an expectation of one large group that is neither very happy or unhappy, as described in **Figure 4.22**.

4) The remaining 32% outside that interval would be divided in half where, 16% would be to the left of one SD and 16% would be to the right of one SD.

Figure 4.22   A Normal Distribution of *Happiness* Scores

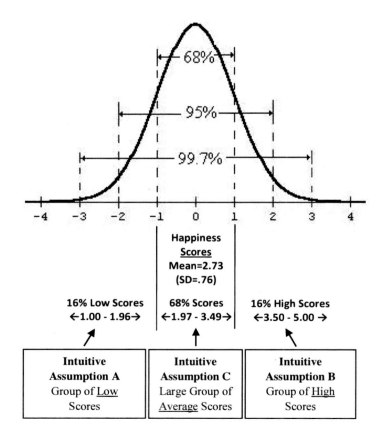

5) The 16% of *Happiness* scores to the right of the one SD mark would be from the value just above 3.49 to the highest scale value (scores 3.50 to 5.00). These scores reflect **Intuitive Assumption B**, which is an expectation of a group of significantly happy study participants, as described in **Figure 4.22**.

6) The 16% of *Happiness* scores to the left side of the one SD mark would be from the lowest scale value to the value just before 1.97 (scores 1.00 to 1.96). These scores reflect **Intuitive Assumption A**, which is an expectation of a group of significantly unhappy study participants, as described in **Figure 4.22**.

Part B: What Are the Threats to a Normal Distribution?

The threats to a distribution of scores being approximately normal include problems regarding skewness, kurtosis, and outlier scores. These threats will be discussed in greater detail below.

Threat 1: Skewness. Skewness is a measure of the asymmetry in a variable distribution. A distribution is symmetric (normally distributed) when the peak is in the center of the distribution and the curve looks the same to the left and right of the center point. *The skewness refers to the lopsidedness or slanting asymmetry.* Skewness is indicated when the peak of the distribution is slanted favorably to either the left (referred to as a positive skew) or right (referred to as a negative skew) hand side of the distribution, as depicted in **Figure 4.23**.

Figure 4.23   The skewness of a distribution of scores

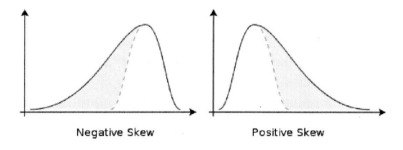

Negative Skew   Positive Skew

Threat 2: Kurtosis. *Kurtosis is a measure concerned with how values are spread out in a normal distribution.* Problems regarding kurtosis occur when the distribution of scores are not properly distributed in the tails of the distribution (the tails are the left hand and right hand slopes of the distribution). Typically, problems occur when the distribution of values are peaked, as in distribution C (leptokurtic) in **Figure 4.24**. This occurs when the values in a distribution are overly clustered around the mean value, leaving the tails empty without many particularly high and low values in the distribution. Alternately, a problem regarding kurtosis might appear like distribution B (platykurtic) in **Figure 4.24**. This occurs when values do not cluster around the mean value score, but are distributed in the tails of the distribution. The normal distributed pattern is in the middle of these two extremes as in distribution A in **Figure 4.24**.

Figure 4.24   Kurtosis in a distribution of scores

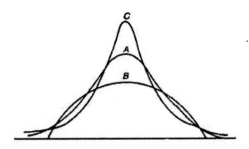

Threat 3: Outlier Scores. Outlier scores are another common threat to distribution being approximately normal. An outlier score is generally a score that is extremely greater than or less than the mean value score within a distribution. These scores are often defined in relation to standard deviations, where a score would be considered an outlier if the value fell two or three standard deviations from the mean score. Outlier scores are important considerations as a small number of outlier scores can easily result in a distribution of scores being non-normal (Barnett & Lewis, 1994). We will discuss how to identify and address these outlier scores later in this section.

Part C: How Do You Assess if Scores are Approximately Normally Distributed?

Here we will discuss the important steps in assessing if the distribution of scores within a continuous variable is approximately normal. In addition, we will apply these techniques to assess whether the values in the dependent variable *Happiness* in our sample study are approximately normally distributed.

Descriptively: Examine the Score Distribution Using Frequency Counts and Charts. The first step involved in assessing whether scores within a continuous variable, such as the dependent variable *Happiness* are normally distributed is to examine the values visually. This can be done in SPSS using the **Frequencies** procedure to produce a histogram with a curve portraying a normal distribution. Go to:

1) **Analyze → Descriptive Statistics → Frequencies**.

2) In the dialogue box that opens, move the *Happiness* from the left hand column into the right hand column using the center arrow (see **Figure 4.25**).

3) Click **Charts**.

Figure 4.25 The dialogue box for the **Frequencies** procedure

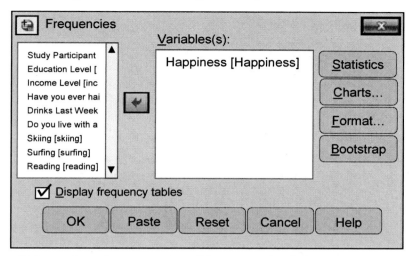

4) In that dialogue box that opens (see **Figure 4.26**) click **Histograms** and the box next to the term **Show normal curve on histogram**.

5) Click **Continue**.

6) Click **Statistics**.

7) Click the box for **Mean**, **Std. deviation**, as well as **Skewness** and **Kurtosis** as presented in **Figure 4.27**.

8) Click **Continue**.

9) Click **OK** or **Paste**.

Figure 4.26 Activate the **histogram** in the frequencies procedure

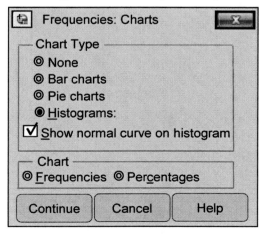

Figure 4.27   Activate the **statistics** in the frequencies procedure

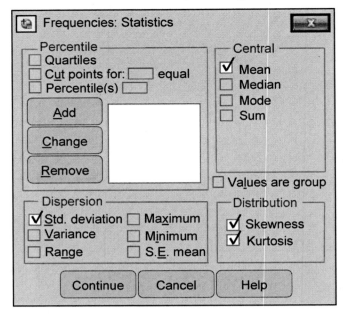

The resulting SPSS output is presented in **SPSS Output 4.4**. Within the output, the first box will indicate the values for the mean, standard deviation, as well as the skewness and kurtosis along with the standard error of each. Later we will discuss how to use these values to assess normality. The output will also include a chart similar to **Figure 4.28**, which we will consider first in terms of assessing normality. This figure presents the histogram graph presenting the distribution of *Happiness* scores with a bell curve shaped line describing a perfectly normal distribution superimposed over the actual distribution.

SPSS Output 4.4

Frequencies

Statistics

Happiness

N	Valid	100
	Missing	0
Mean		2.7340
Std. Deviation		.75535
Skewness		.196
Std. Error of Skewness		.241
Kurtosis		-.497
Std. Error of Kurtosis		.478

Figure 4.28 Histogram of *Happiness* scores with a normal curve

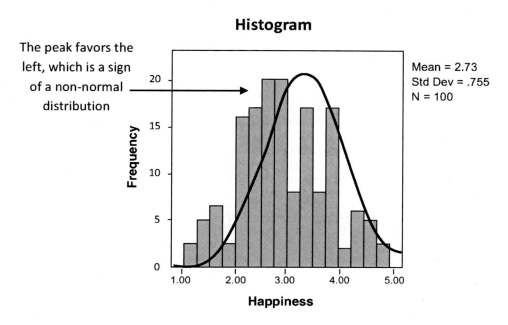

Within **Figure 4.28**, we see the distribution of *Happiness* scores (formed by the columns) approximates a bell curve shape. For example, there is an identifiable peak and a number of scores in both tails. However, there are some spaces in the tails and in the center not populated by scores. Also, there are scores above the normal curve. Furthermore, although there is a clear peak, the center of the peak is skewed to the left (positively skewed).

We know that optimally the peak would be in the direct center of the distribution, but here the peak is a little off. Is this a big problem, little problem, or no problem at all? What we need now is a specific criterion to identify whether the distribution of scores approximates a normal distribution. *Therefore, we will now apply specific statistical criteria to assess if a distribution is approximately normal* through statistical testing (*The Kolmogorov–Smirnov* and *Shapiro–Wilk Statistical Tests*), as well as considering the ratio of the value of skewness and kurtosis relative to the standard error of each.

Statistically (1): The Kolmogorov–Smirnov and Shapiro–Wilk Statistical Tests. The Kolmogorov-Smirnov Test (Kolmogorov, 1933; Smirnov, 1948) and the Shapiro-Wilk (Shapiro & Wilk, 1965) are two well-known statistical procedures used to test the level of normality reflected in a distribution of values. The tests are each fairly rigorous. The Kolmogorov–Smirnov and Shapiro–Wilk tests compare the distribution of scores of a variable to a normally distributed set of scores with the same mean and standard deviation. When either test fails to achieve statistical significance ($p>.05$), this indicates that the distribution of scores within the variable is not significantly different from a normal distribution. In other words, the distribution of scores is approximately normal. However, if either test does achieve statistical significance ($p<.05$), this indicates that the distribution of scores is significantly different from a normal distribution. It is said that the Shapiro-Wilk Test is more appropriate for small sample sizes (<50), but it is appropriate for large sample sizes too (as large as 2000). We will perform both tests toward assessing our dependent variable *Happiness*. To perform these tests in SPSS:

1) Click **Analyze → Descriptive Statistics → Explore**.

2) In the dialogue box that opens, transfer the variable (*Happiness*) to be tested for normality from the column on the left into the cell marked **Dependent List** on the right as presented in **Figure 4.29**.

3) Click **Plots**.

4) In the dialogue box that opens, click **Normality plots with tests** (see **Figure 4.30**).

5) Click **Continue**.

6) Click **OK** or **Paste**.

Figure 4.29   The dialogue box for the **Explore** procedure

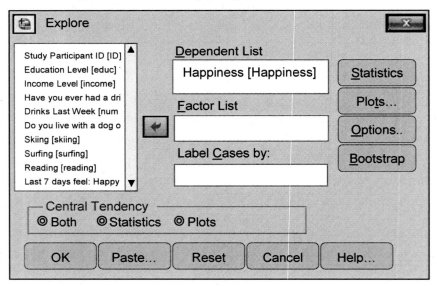

SPSS Output 4.5 presents the box labeled **Tests for Normality**, which will appear within this output. Under the column labeled **Kolmogorov-Smirnov** we see the value of the statistic (Statistic=.110), the degrees of freedom (df=100), and the level of

probability (Sig.=.004). Under the column labeled **Shapiro-Wilk** we also see the value of the statistic (Statistic=.979), the degrees of freedom (df=100), and the level of probability (Sig.=.105). For either test, when the value in the **Sig.** (which is the value of probability) box is .05 or less, the distribution of scores is significantly non-normal.

Figure 4.30   Activate the **plots** in the explore procedure

Check *Normality* *plots with tests*

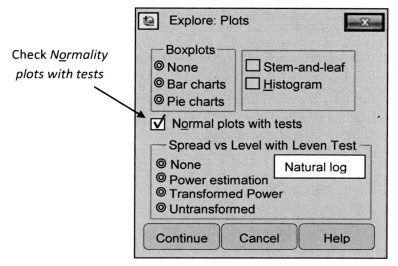

In terms of the current results, we see the value of **Sig.** for the **Kolmogorov-Smirnov** test is .004, which is below .05 ($p<.05$), which indicates that the distribution of scores for the dependent variable *Happiness* is significantly non-normal (as per this test). However, we also see the results of the **Shapiro-Wilk** test indicate that the results are not statistically significant, as **Sig.** was greater than .05 ($p=.105$). Recall, we mentioned earlier that the **Shapiro-Wilk** test is more appropriate for smaller samples, which would apply to our sample, as 100 study participants is relatively small.

SPSS Output 4.5

Tests of Normality

	Kolmogorov-Smirnov[a]			Shapiro-Wilk		
	Statistic	df	Sig.	Statistic	df	Sig.
Happiness	.110	100	.004	.979	100	.105

Non-normal distribution indicated as **Sig.** is less than .05 ($p<.05$)

Approximately normal distribution indicated as **Sig.** is greater than .05 ($p>.05$)

Thus, it might be quite appropriate to use the results of the **Shapiro-Wilk** test and discard the **Kolmogorov-Smirnov** as the **Shapiro-Wilk** is more appropriate for our sample size. However, we also might like to observe why the **Kolmogorov-Smirnov** indicated a non-normal distribution. Toward this, recall our initial visual inspection of the distribution of scores in the histogram that seemed to suggest traits of a non-normal distribution such as a slight skew. We can then speculate that this skew may be related to the **Kolmogorov-Smirnov** test being statistically significant.

Although the **Kolmogorov-Smirnov** and **Shapiro-Wilk** statistical tests may be convenient and clear, each test is quick to identify a distribution of scores as non-normally distributed at a significant level (these statistical tests are rather sensitive). Therefore, it is often useful to be aware of other methods of assessing normality besides these two statistical tests, such as considering the ratio of the skewness and kurtosis to the standard error of each.

Statistically (2): The Ratio of Skewness and Kurtosis to the Standard Error of Each. Recall the earlier **frequencies** procedure that produced **Statistics** for the distribution of *Happiness* scores that included values for the skewness and kurtosis for this variable, as well as the standard error of each value. The normality of a distribution of scores can be examined by considering the ratio between the skewness and the standard error of the skewness, as well as the kurtosis and the standard error of the kurtosis (Miles & Shevlin, 2001). Specifically, by dividing either value by the respective standard error of that statistic, a standard score is produced that represents the ratio between the two values. In general, if the resulting ratio is two or less, the distribution of scores is assumed to be approximately normally distributed. Alternately, you can also multiply the standard error by '2' and observe if the value of the respective skewness or kurtosis is less than the product. For example, to compute these numbers for the skewness of the dependent variable *Happiness*, using the values presented in **SPSS Output 4.6**:

1) Multiply the standard (Std.) error of skewness (.241) by two, which is .482.

2) Note that the value of the skewness (.196) is less than twice the standard error (.482) of the skewness, which indicates an approximately normal distribution.

3) To compute the exact ratio as mentioned initially, divide the skewness (.196) by the standard error of the skewness (.241), which is .81.

4) The ratio between the skewness and standard error of the skewness is .81, which is below two, indicating an approximately normal distribution.

To compute these numbers for the kurtosis of the dependent variable *Happiness*, using the values presented in **SPSS Output 4.6**:

1) Multiply the standard (Std.) error of kurtosis (.478) by two, which is .956.

2) Note that the value of the kurtosis (-.497) is less than twice the standard error (.956) of the kurtosis, which indicates an approximately normal distribution.

3) To compute the exact ratio as mentioned initially, divide the kurtosis (-.497) by the standard error of the kurtosis (.478), which is 1.04.

4) The ratio between the kurtosis and the standard error of kurtosis is 1.04, which is below two, indicating an approximately normal distribution.

SPSS Output 4.6

Statistics

Happiness

N	Valid	100
	Missing	0
Mean		2.7340
Std. Deviation		.75535
Skewness		.196
Std. Error of Skewness		.241
Kurtosis		-.497
Std. Error of Kurtosis		.478

Skew (.196) is less than twice the standard error (.241 x 2 = .482)

Kurtosis (-.497) is less than twice the standard error (.478 x 2 = .956)

Lastly, graphs are often used to assess the normality of a distribution of scores. For example, a commonly used graph to this end is the Q-Q Plot (see **Figure 4.31**).

Figure 4.31   The Q-Q plot describing the normal distribution

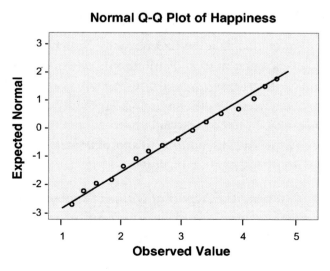

Normal Q-Q Plot of Happiness

The Q-Q plot can be produced along with the **Kolmogorov-Smirnov** and **Shapiro-Wilk** tests through the clicking **Normality plots with tests** (At **Analyze →Descriptive Statistics → Explore→ Plots**).This plot indicates a normal distribution of scores when data points are close to the diagonal line within the graph. As we can see from the Q-Q Plot presented in **Figure 4.31** based on the dependent variable *Happiness*, the data seem to be closely assembled around the straight diagonal line indicating a normal distribution.

Part D: How Do You Address Problems Related to a Non-Normal Distribution?

There are several methods to modify a non-normal distribution of scores to make the values better approximate normality. We will consider two of the most commonly used methods as options. Specifically, we will first examine how to modify a non-normal distribution by assessing the impact that outlier scores have on making a distribution non-normal. Second, we will review the popular practice of data transformations.

Option 1: Assessing the Impact of Outlier Scores (Steps 1-4). An outlier score is a value that is outside the normal range of the other scores in the distribution. In other words, one could say scores are called outliers because the values *lie* far *out* from the area surrounding the mean score. Recall how in our examination of **Figure 4.22,** we categorized scores into high and low *Happiness* scores if values were more than one SD from the mean score. Valid as that might be, an outlier score is typically as least two or three SDs outside the mean score. The outlier score tends to be extremely different from the mean score and thus often makes a distribution significantly non-normal. Subsequently, when you need to assess the cause of a non-normal distribution, a good first move is to assess if an outlier score(s) is influencing the distribution.

As an exercise, in light of the fact that the **Kolmogorov-Smirnov** was statistically significant (which indicates a non-normal distribution), let's modify the distribution of scores within the dependent variable *Happiness* to be more normally distributed by excluding the outlier scores from the distribution. After modifying the variable, let's then repeat the **Kolmogorov-Smirnov** test and observe if the results are no longer statistically significant (which would indicate an approximately normal distribution). To do this let's follow **Steps 1, 2, 3,** and **4** below.

STEP 1 (of *Option 1: Assessing the Impact of Outlier Scores*): **Identify the Outlier Score(s).** There are several straightforward ways to identify the outlier score(s) in a distribution. Here we will look at using a boxplot, as well as the raw mean and standard deviation scores to do so.

Method 1 to Identify Outlier Scores: The Boxplot. A commonly used tool to identify an outlier score(s) is the boxplot, which can be produced in several ways in SPSS, including the **Explore** function at:

1) **Analyze → Descriptive Statistics → Explore.**

2) To identify outlier scores, move the applicable variable, e.g., *Happiness*, from the column on the left into the cell on the right labeled **Dependent List.**

3) The toggle button in the lower left hand corner must be checked for either **Plots** or **Both** for the boxplot figure to be produced (as presented in **Figure 4.29**).

4) Next, click **OK** or **Paste**.

The **Explore** function will produce the boxplot presented in **Figure 4.32**. If there are outlier scores within the distribution of scores, each score will appear as a small circle with a corresponding number. For example, note in the boxplot presenting *Happiness* scores, there are two small circles above the central figure, which are the high *Happiness* outlier scores. Likewise, there are two small circles below the central figure, which are the low *Happiness* outlier scores.

The numbers next to the circles indicate where the case with the outlier score is located in reference to the blue column of numbers on the extreme left hand side of the **Data View** section of the SPSS database. Next, we will review how to find the outlier scores indicated on the boxplot, within the SPSS dataset, so these scores can be included and excluded from analysis.

Figure 4.32 Identifying outlier scores in a boxplot display

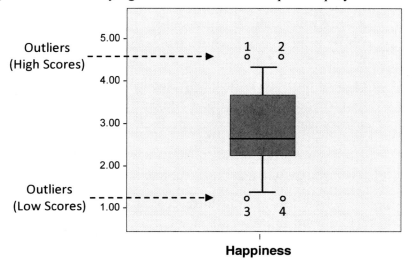

Happiness

For a better illustration, **Figure 4.33** presents the **Data View** portion of the SPSS database that describes how to identify the outlier score in the raw data according to the numbers indicated in the boxplot. The first two lines in the blue column to the left labeled '1' and '2' (in the raw data) correspond to the numbers '1' and '2' next to the circles in the boxplot. These are the high outlier scores that correspond to the study participant ID 13 and ID 15, who each have a score of 4.40.

The next two lines in the blue column to the left labeled '3' and '4' (in the raw data) correspond to the numbers '3' and '4' next to the circles in the boxplot. These are the low outlier scores, that correspond with study participant ID 39 and ID 88, who each have a score of 1.20.

Figure 4.33   Viewing the outlier *Happiness* scores

	ID	Happiness	
1	13.00	4.40	High outlier scores
2	15.00	4.40	(On boxplot #s 3 & 4)
3	39.00	1.20	Low outlier scores
4	88.00	1.20	(On boxplot #s 1 & 2)

Numbers on the *boxplot* (1, 2, 3, & 4) correspond
with the numbers in the *column shaded blue*

Method 2 to Identify Outlier Scores: The Raw Mean and Standard Deviation Values. As mentioned earlier, we can use the raw mean score and standard deviation values to identify outlier scores. Specifically, it would be appropriate to define an outlier score as a score greater than two units of standard deviation from the mean score. For example, the mean score for *Happiness* is 2.73, while one unit of standard deviation is .76. If we define an outlier score as a score more than two units of standard deviation from the mean score, the outlier score must be more than 1.52 (2 units of standard deviation = 2*.76=1.52) greater than or less than the mean score. Specifically:

1) A high outlier score is a score greater than the mean score of 2.73 plus 1.52 (2 units of SD), which is 4.25 (we will call this computation #1).

2) A low outlier score is a score less than the mean score 2.73 minus 1.52 (2 units of SD), which is 1.21 (we will call this computation #2).

Using these computations, we can now state the *intervals within the range of Happiness scores that the high and low outlier scores would fall within.*

Interval 1: High Outlier Scores for Happiness. From computation #1, we know a *Happiness* score greater than 4.25 ($\geq$4.26) is an outlier score. To compute this interval, we also need to state the highest possible score for the *Happiness* scale, which is 5.00. Thus, the interval for a high outlier score is <u>4.26 to 5.00</u>.

Interval 2: Low Outlier Scores for Happiness. From computation #2, we know a *Happiness* score less than 1.21 is an outlier score ($\leq$ 1.20) is an outlier score. To compute this interval, we also need to state the lowest possible score for the *Happiness* scale, which is 1.00. Thus, the interval for a low outlier *Happiness* score is <u>1.00 to 1.20</u>. We can then observe the number of outlier scores that fall within the two intervals through conducing a **frequency analysis** of the variable *Happiness*. In SPSS go to:

1) **Analyze→ Descriptive Statistics→ Frequencies**.

2) In the dialogue box that opens (see **Figure 4.25**), move the variable *Happiness* from the left hand column into the right hand column.

3) Click **OK** or **Paste**.

SPSS Output 4.7

Happiness

	Frequency	Percent	Valid Percent	Cumulative Percent	
1.20	**2**	**2.0**	**2.0**	**2.0**	
1.40	3	3.0	3.0	5.0	Outliers Scores (Low):
1.60	4	4.0	4.0	9.0	<1.21
1.80	2	2.0	2.0	11.0	
2.00	9	9.0	9.0	20.0	
2.20	10	10.0	10.0	30.0	
2.40	12	12.0	12.0	42.0	
2.60	12	12.0	12.0	54.0	
2.80	5	5.0	5.0	59.0	
3.00	10	10.0	10.0	69.0	
3.20	5	5.0	5.0	74.0	
3.40	10	10.0	10.0	84.0	
3.60	6	6.0	6.0	90.0	
3.80	1	1.0	1.0	91.0	Outliers Scores (High):
4.00	4	4.0	4.0	95.0	>4.25
4.20	3	3.0	3.0	98.0	
4.40	**2**	**2.0**	**2.0**	**100.0**	
Total	100	100.0	100.0		

The **SPSS Output 4.7** presents the **frequency table** included in the resulting statistical output. Within the frequency table, we can determine how many scores fall within the intervals reflecting high outlier scores (from 4.26 to 5.00) and low outlier scores (from 1.00 to 1.20). Subsequently, we see this distribution reveals two high outlier scores of 4.40 (within the interval 4.26 to 5.00), and two low outlier scores of 1.20 (within the interval 1.00 to 1.20). These values are highlighted within the frequency table. We could also imagine where the outlier *Happiness* scores would fall within the bell curve describing a normal distribution. **Figure 4.34** presents a normal distribution bell curve with the outlier *Happiness* scores superimposed. The callout balloons identify where the outlier scores noted in the frequency table presented in **SPSS Output 4.7** would fall within this bell curve distribution.

Figure 4.34   Outlier *Happiness* scores (>2 SD) in the distribution

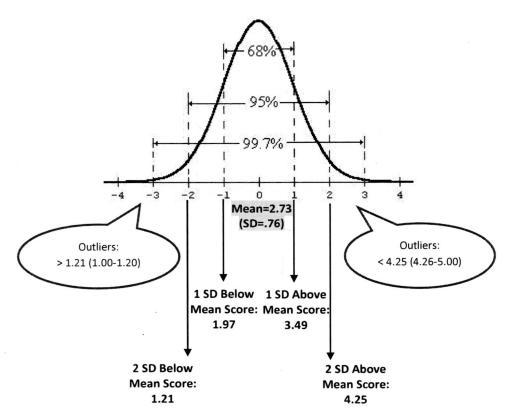

Within the figure, the scores that reflect one and two units of standard deviation are noted beneath the figure. Notice that the extreme high *Happiness* scores of 4.40 are indicated by the callout to the right of the two standard deviation mark above the mean. Additionally, the extreme low *Happiness* scores of 1.20 are indicated by the callout to the left of the two standard deviation mark below the mean.

STEP 2 (of *Option 1: Assessing the Impact of Outlier Scores*): Test the Distribution of Scores for Normality Without the Outlier Score(s). The next step involves examining the distribution of scores for normality once the outlier scores have been excluded from the analysis. In order to do this, you will need to perform three tasks.

1) First, you must **create a filter variable** to identify the outlier scores.

2) Second, use the **select cases** function in SPSS to exclude the outlier scores from analysis.

3) Third, test the distribution for normality with the **outlier scores excluded** from analysis.

Accordingly, STEP 2 will be presented as three separate TASKS in this section.

STEP 2, TASK 1: *Create a Filter Variable to Identify the Outlier Scores.* The first step in this process is to create a filter variable to identify the outlier scores in the distribution. You can identify the outlier scores through recoding the variable reflecting *Happiness* according to which variable scores are within and outside the two standard deviation intervals indicating outlier scores. In SPSS to perform a variable recode go to:

1) **Transform → Recode into Different Variables**.

2) In the dialogue box that opens, move the variable reflecting *Happiness* from the left hand column into the center column (labeled **Numeric Variable → Output Variable**).

3) Within the cell labeled **Name:** on the right hand side of the dialogue box, enter a name for the new variable (such as the name *HappyOUTLIER* entered below)

4) Enter a variable description in the cell marked **Label** (such as *Happiness Outlier Score*).

5) Click the button marked **Change** to move the new variable name into the center column, as presented in **Figure 4.35**.

6) Click **Old and New Values**.

7) In the **Old and New Values** dialogue box, under the column on the left marked **Old Value**, click the toggle button for **Range**, and enter the interval for the high outlier scores 4.26-5.00.

8) Under the column marked **New Value** on the right, enter a '1' (1=Outlier Scores) and then click the button underneath marked **Add**.

Figure 4.35   Recode Into Different Variables: *Happiness* into *HappyOUTLIER*

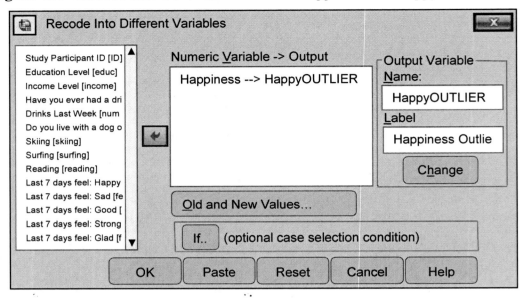

Figure 4.36   Old and New Values: Recoding *Happiness* into *HappyOutlier*

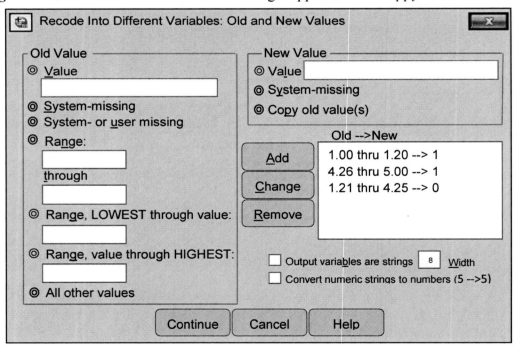

9) Under the column on the left marked **Old Value**, click the toggle button for **Ra̲nge**, and enter the interval for the low outlier scores 1.00-1.20.

10) Under the column marked **New Value** on the right, enter a '1' (1=Outlier Scores) and click the button underneath marked **A̲dd**.

11) Under the column on the left marked **Old Value**, click the toggle button for **Range**, and enter the interval for scores that are not outlier scores 1.21-4.25.

12) Then under the column marked **New Value** on the right, enter a '0' (0=Not Outlier Scores) and click the button underneath marked **Add**.

13) The dialogue box will look similar to **Figure 4.36**.

14) Click **Continue**.

15) Click **OK** or **Paste**.

If you go to the end of the variable list in either the **Data View** or **Variable View** a new variable should have been created in the SPSS database named *HappyOUTLIER* where a '1' indicates an outlier score and a '0' indicates a score that is not an outlier. It is a good idea to verify that the new variable correctly identifies the outlier scores by checking the variables in the raw data. Do this by:

1) Go into the **Data View** portion of the SPSS database.

2) Locate the new variable *HappyOUTLIER* at the end of the list of variables.

3) Line up the original variable *Happiness* next to the new variable *HappyOUTLIER*.

4) Notice as presented in **Figure 4.37**, that there is:

    a. A '1' in the variable *HappyOUTLIER* were outlier scores (1.20, 1.20, 4.40, and 4.40) are present in the variable *Happiness*.

    b. A '0' in the variable *HappyOUTLIER* were non-outlier scores (other than 1.20, 1.20, 4.40, and 4.40) are present in the variable *Happiness*.

Figure 4.37 Comparing *Happiness* and *HappyOUTLIER* values

	Happiness	HappyOUTLIER	
	4.40	1.00	
	4.40	1.00	
4.40 and 1.20 are outlier scores	1.20	1.00	1.00 = *Yes* (Is outlier score)
	1.20	1.00	
4.20 is not an outlier score	4.20	0.00	0.00 = *No* (Not outlier score)
	4.20	0.00	

As a second check, also run a **Frequencies** procedure:

1) Click **Analyze → Descriptive Statistics → Frequencies**.

2) Move the variable *HappyOUTLIER* from the left hand column into the right hand column.

3) Click **OK** or **Paste**.

SPSS Output 4.8 presents the frequency table within the statistical output. We know there were four outlier scores, so the frequency table should describe a count of '4' for the response 1.00 (outlier scores) and '0' for the remaining 96 scores (not outlier scores). The output reflects these numbers suggesting our coding was successful.

SPSS Output 4.8

Happiness Outlier Scores

		Frequency	Percent	Valid Percent	Cumulative Percent
Valid	.00 (=No)	96	69.0	96.0	69.0
	1.00 (=Yes)	4	4.0	4.0	4.0
	Total	100	100.0	100.0	

STEP 2, TASK 2: *Use the Select Cases Function to Exclude the Outlier Scores from Analysis.* Next, use the **Select Cases** function in SPSS to exclude the outlier scores from analysis. To do this in SPSS follow the path:

1) **Data → Select Cases**.

2) When the dialogue box opens, in the top center click *If condition satisfied*.

3) Click the button below that reads *If....* (see **Figure 4.38**).

Figure 4.38   Select Cases function (top of dialogue box only)

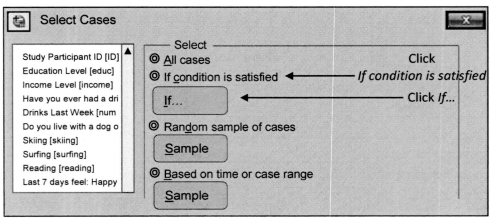

4) Within the new dialogue box use the button in the top center to bring the filter variable *HappyOUTLIER* from the left into right hand column.

5) Then use the center keypad to enter '=0' after the term to tell SPSS to only include cases with a '0' in the analysis regarding the variable *HappyOUTLIER* as presented in **Figure 4.39** (thereby excluding the cases marked '1').

Figure 4.39 The **If** function within the select cases (top of dialogue box only)

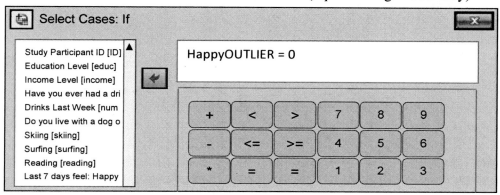

6) Click **Continue**.

7) Click **OK** or **Paste**.

Hash marks. When cases are deselected from analysis, *hash marks* will appear in the **Data View** portion of the SPSS database next to the cases excluded from analysis. Therefore, after selecting/deselecting cases make sure to go back to the **Data View**

Figure 4.40 Hash marks indicating outlier scores are excluded from analysis

Hash marks reflect cases excluded from analysis (Where HappyOUTLIER=1)

	ID	Happiness	HappyOUTLIER	
1	13.00	4.40	1.00	(1 = Yes)
2	15.00	4.40	1.00	
3	39.00	1.20	1.00	
4	88.00	1.20	1.00	
5	10.00	4.20	0.00	(0 = No)
6	11.00	4.20	0.00	

No hash marks reflect cases included in analysis (Where HappyOUTLIER=0)

section of the SPSS database to check that the hash marks (the diagonal black lines) only appear (over the number in the blue left hand column) for the cases targeted for exclusion. For example, in our analysis, we would expect hash marks over the cases where the variable *HappyOUTLIER* has a '1' entered (indicating an outlier score) as presented in the **Figure 4.40**.

STEP 2, TASK 3: *Test the Distribution of Scores for Normality with the Outlier Scores Excluded.* Now that the outlier scores have been excluded from analysis using the **select cases** function, we can re-test to reveal if the distribution of scores is approximately normal without the influence of those extreme scores. To do this, compute the statistical test run initially, the **Kolmogorov-Smirnov** and **Shapiro-Wilk** tests and note if both are not statistically significant, which would indicate that the distribution of scores is approximately normal with the outlier scores excluded.

SPSS Output 4.9 presents the **Kolmogorov-Smirnov** and **Shapiro-Wilk** test conducted with the outlier scores excluded. Recall in our original analysis, the distribution of scores for the dependent variable *Happiness* was described as significantly non-normal as per the **Kolmogorov-Smirnov** test, as the significance level was less than .05 ($p<.05$). However, we can now see that the result of the **Kolmogorov-Smirnov** test indicates a significance level of .150. Thus, as the level of significance is greater than .05 ($p>.05$), we can assume that this test now indicates that the distribution of *Happiness* scores now approximates a normal distribution. Thus, it seems that the four outlier scores may have been causing the distribution of *Happiness* scores to be relatively non-normal as per the **Kolmogorov-Smirnov** test.

SPSS Output 4.9

Tests of Normality

	Kolmogorov-Smirnov[a]			Shapiro-Wilk		
	Statistic	df	Sig.	Statistic	df	Sig.
Happiness	.072	96	.150	.977	96	.091

Approximately normal: $p>.05$

STEP 3 (of *Option 1: Assessing the Impact of Outlier Scores***): Assess if Removing the Outlier Score(s) Supported a Normal Distribution of Scores.** Next, compare the findings of the initial tests of normality with the second tests of normality where the outlier scores have been removed. If the initial analysis indicates a non-normal distribution (e.g., **Kolmogorov-Smirnov** = $p<.05$), but the second indicated an approximately normal distribution (e.g., **Kolmogorov-Smirnov** = $p>.05$), the analysis likely indicates that the removal of outlier scores is supporting a normal distribution.

This is an important consideration as the removal of outlier scores can have several effects, including having no effect on normality or a significant effect on normality, as is the case in our analysis. Thus, if we are trying to enhance normality, we need to decide if this method is useful for our dataset or not. Obviously, if the method is not useful, we would need to look for another tool to enhance normality, such as the data transformations discussed in the next section: *Option 2: Data Transformations*.

However, in the case of our data, removing the four outlier scores promoted the distribution of *Happiness* scores in being approximately normal. Thus, the question now becomes, how can we use this information to satisfy the test assumption of a normal distribution in the parametric test of multiple linear regression we are employing? For an answer see Step 4 below.

STEP 4 (of *Option 1: Assessing the Impact of Outlier Scores*): Find Evidence of Robustness in the Parametric Testing (e.g., multiple linear regression) by Removing and Reinserting the Outlier Scores from the Analysis. In order to make use of your discovery that removing the outlier scores from the distribution supports an acceptable level of normality, you must:

1) First assess if the inferential tests (in this case a multiple linear regression analysis model) are performing in a robust manner. Recall, a parametric test such as multiple linear regression analysis is said to be "robust" when a test assumption (such as a normal distribution of scores) is violated but the statistical test still performs well in spite of the violation(s) (van Belle, 2002).

2) If you can provide evidence that a statistical test is acting in a robust manner, you can allow the violation of the test assumption (e.g., a non-normal distribution) to exist within the analysis (but remember to report this exists when reporting results).

3) To indicate if a statistical test is performing in a robust nature conduct the data analysis (e.g., multiple linear regression) using the original variable with the non-normal distribution of scores.

4) Conduct the analysis (e.g., multiple linear regression) identically, except this time exclude the outlier scores from the analysis.

5) Note if the results of the data analysis (e.g., the two regression models) differ significantly ($p<.05$) when the outlier scores are included and excluded from the analysis.

To illustrate this practice of comparing the two test results, let's use our analysis as an example. **SPSS Output 4.10** presents the results of the first multiple linear regression test that uses the normally distributed version of *Happiness* scores (using 96 scores, with the 4 outlier scores excluded). Please note, within the context of the full multiple linear regression model, all predictors evidenced a statistically significant ($p<.05$) relationship with the dependent variable *Happiness*, as indicated by all the values in the final column to the right labeled **Sig** being below the value .05.

SPSS Output 4.10

Coefficients[a]

Model	Unstandardized Coefficients		Standardized Coefficients	t	Sig.
	B	Std. Error	Beta (β)		
Years of Education	.06	.02	.212	2.6	.011
Income Level	.01	.00	.481	5.8	.000
Live with a dog or a cat?	-.34	.11	-.242	-2.9	.004

All relationships are significant ($p<.05$) when **outlier scores are excluded (n=96)**

SPSS Output 4.11 presents the results of the second multiple linear regression test that uses the non-normally distributed version of *Happiness* scores (using 100 scores, with the 4 outlier scores included). Please note, within the context of the full multiple linear regression model, all predictors evidenced a statistically significant ($p<.05$) relationship with the dependent variable *Happiness*, as indicated by all the values in the final column to the right labeled **Sig** being below the value .05.

SPSS Output 4.11

Coefficients[a]

Model	Unstandardized Coefficients		Standardized Coefficients	t	Sig.
	B	Std. Error	Beta (β)		
Years of Education	.05	.02	.18	2.3	.027
Income Level	.01	.00	.48	6.0	.000
Live with a dog or a cat?	-.42	.12	-.28	-3.5	.001

All relationships are significant ($p<.05$) when **outlier scores are included (n=100)**

So we see when we compare the regression models that the statistically significant ($p<.05$) relationships between the predictors and the dependent variable do not change when the outlier scores are included or excluded from analysis. Thus, the test of multiple linear regression is performing in a "robust" manner, where the violation of the test assumption of an approximately normal distribution does not seem to be compromising the study results. Subsequently, one might consider using the version of the *Happiness* variable with the somewhat non-normally distribution of scores in the analysis. However, it is important to describe that this was done in the methods section of the research report or manuscript in which this analysis is presented.

However, if we conducted these two multiple linear regression analyses and found that the relationships between the predictor variables and dependent variable differed significantly (in terms of statistically significance, directionality, effect size, etc.), we would not indicate the test was performing in a robust manner. In that instance we would need another strategy for addressing the non-normal distribution of scores aside from removing and reinserting the outlier scores from the analysis.

Removing the Outlier Scores from the Analysis Permanently. Lastly, when an outlier score(s) is identified as the factor impacting the normality of a distribution of scores, many analysts *remove the outlier score* by deleting the case with that score from analysis. However, it is important to realize that there are guidelines that should be observed for permanently deleting a case from the analysis (Field, 2005). Specifically, an analyst often has other reasons for deleting the outlier score other than it is messing up his or her day by making a distribution of scores non-normal.

For example, an analyst might delete a case with the outlier score from analysis if the case seems not to be from the population under study. For example, the target population of our sample study examining *Happiness* might be people that are not taking medication (e.g., anti-depressants), because taking medication might enhance *Happiness*. Thus, if there was some extremely high score for *Happiness* that was only possible through medication (obviously hypothetical), an analyst would be justified in removing that outlier score from the analysis as the extreme score suggests that this study participant is not from the target population.

Option 2: Data Transformations. Another means of addressing a non-normal distribution (including reducing the impact of outliers) is to perform a data transformation (see the three articles for Bland & Altman, 1996). The first thing you should realize about data transformations is that when the data are transformed, the values associated with the distribution (e.g., the mean and standard deviation) of scores are altered. Thus, when transforming the data, you might gain a normality regarding the distribution, but you are losing specific characteristics of the raw data. However, because all the scores are transformed in the same way, the relationships between the variables will remain the same, which is essential.

A data transformation applies a deterministic mathematical function (e.g., multiplication) to each score to make the distribution of scores better suited for use in statistical testing (e.g., multiple linear regression). There are different types of data transformations to address different challenges that make distribution of scores non-normal. For example, there are different methods to address a distribution with a significantly positive or negative skew, as well as when data are too peaked or flat (challenges related to kurtosis). For example, we will now discuss how to use a **Log Transformation** to help make a distribution of scores more normal.

Log Transformation. A **Log Transformation** can be used to address a distribution of scores that is non-normal due to a significantly positive skew (to the left). Keep in mind if you have a distribution of scores that is significantly negatively skewed (to the right), you can reverse code the scores, so the significant skew is reversed in the other direction to be positive (to the left), at which point you can apply the **Log Transformation** to that distribution.

Taking a *logarithm* (Log) of a set of numbers move the numbers to the right tail (right hand side) of the distribution. Thus, if you examine the histogram chart presented earlier (which presents the bell-curve distribution of scores) and note the peak of the distribution favors the left, the **Log Transformation** might be a good method of making the distribution approximately normal by reducing the positive skew.

NOTE: You can't get a log value using a zero or negative score or value. Thus, if there is a zero or negative value(s) in the distribution of scores, you must make all values positive prior to performing the **Log Transformation**. To do this, add a constant number to all scores, so the smallest number in the distribution is made positive. For example, if the smallest number in a distribution is -3, add a 4 (using the Transform function in SPSS: **Transform→ Compute Variable**) to each score in the distribution.

You may recall when we assessed *Happiness* scores descriptively, our histogram with a normal curve estimation indicated the distribution was slightly skewed to the left (positively skewed). Therefore, we will conduct a **Log Transformation** of our distribution of *Happiness* scores as a means of increasing the normality of this distribution by reducing the positive skew. To conduct a **Log Transformation** of *Happiness* scores (the variable *Happiness*) in SPSS go to:

1) **Transform→ Compute Variable.**

2) In the dialogue that opens, in the center of the right hand side of the box labeled **Function group**.

3) In the box **Functions and Special Variables**, highlight the term **Lg10**.

4) Use the arrow (pointing up) to the left of that **Functions and Special Variables** box, move the term **Lg10** into the top box marked **Numeric Expression**.

5) Move the variable to be transformed (in our case *Happiness*) from the left hand column in between the parentheses within the term **Lg10**.

6) Then in the upper left hand corner under the term **Target Variable** enter the name for our new transformed variable. For our purposes we have named the **Log Transformed** version of the variable *Happiness* to be *HappyLOG*.

7) The dialogue will then look similar to **Figure 4.41**.

8) Click either **OK** or **Paste**.

Figure 4.41   Computing the **Log Odds** function (top of dialogue box only)

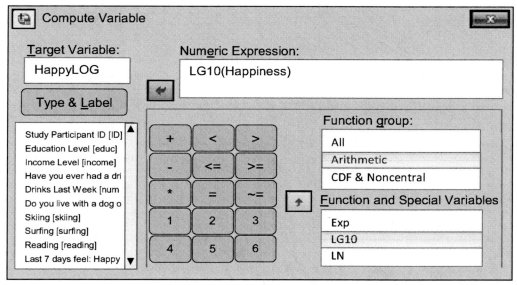

Next, conduct a frequencies procedure using the transformed variable *HappyLOG* to generate a histogram presenting the modified distribution of *Happiness* scores. In SPSS, go to:

1) **Analyze → Descriptive Statistics → Frequencies**.

2) In the dialogue box that opens (see **Figure 4.25**), move the *HappyLOG* from the left hand column into the right hand column using the center arrow.

3) Click **Charts**.

4) In that dialogue box that opens click **Histograms** and the box next to the term **Show normal curve on histogram** (as in **Figure 4.26**).

5) Click **Continue**.

6) Click **OK** or **Paste**.

Figure 4.42 presents the histogram presenting the log odds modified distribution of *Happiness* scores (*HappyLOG*). Recall, the original non-transformed distribution of *Happiness* scores had certain features that reflected a non-normal distribution. For example, the peak in the original distribution of *Happiness* scores favored the left hand side of the distribution (positive skew). As mentioned earlier, the log odds transformation is designed to address this challenge by pushing scores to the right to compensate for the non-normal skew. If we observe the data transformed version of *Happiness* scores, which is the variable *HappyLOG* (**Figure 4.42**), we see the scores are pushed to the right to better approximate a more normal distribution.

Figure 4.42   The log odds transformed distribution of *Happiness* scores

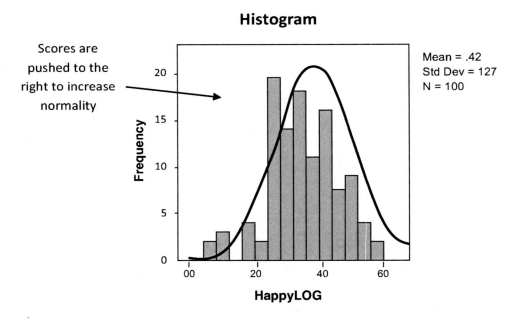

As mentioned earlier, there are several different types of data transformations used to address different challenges related to a non-normal distribution, including the **Square Root Transformation**, which brings outlier scores toward the center of the distribution. Another common procedure used to estimate a continuous variable is an **Inverse (Reciprocal) Transformation**. However, at times, even various well thought out data transformation will not result in an approximately normal distribution of scores. Here analysts often consider more extreme methods such as changing the actual values of outlier scores toward making a distribution more normal. For example, an outlier score might be changed to be just above the next highest score. However, at this point you must wonder which is worse, having a non-normal distribution that violates a test assumption or changing an actual raw score in the distribution.

When the Regression Model Is Robust, Consider Using the Non-Transformed Version. Recall, in the prior section we discussed that once the outliers were removed we could retest the regression model without the outliers to observe if the statistical results (e.g., predictor/Dependent Variable relationships are statistically significant at the .05 level) were similar when the test assumption was met (i.e., no outlier scores) or not met (i.e., outlier scores in the distribution). Then we mentioned that if this occurs, the model may be showing signs of robustness in reference to the violation of the test assumption. When this occurs we might consider using the variable in the original form to accommodate data fidelity.

If the results are similar in a regression model when using the transformed (to better approximate a normal distribution) and original version of a variable, we might want to consider this same method here. Specifically, if a variable is transformed to better approximate a normal distribution and the regression model yields similar statistical results (e.g., predictor/Dependent Variable relationships are statistically significant at the .05 level) when the test assumption are met (i.e., the normal distribution resulting from the transformation) or not met (i.e., the original variable form where the distribution is non-normal), we might consider using the variable in the original form to accommodate data fidelity.

4.4.2.2 Multicollinearity

Multicollinearity is a necessary check when conducting a regression model with multiple predictors. Again, before a researcher can have confidence in the results of a regression model, an assessment must be made evidencing that a problematic degree of multicollinearity does not exist between predictor variables. In this section we will review how to assess multicollinearity in the following three parts:

Part A: What is Multicollinearity?

Part B: How Do You Assess if Multicollinearity Exists?

Statistically (1): Correlations

Statistically (2): Collinearity Diagnostics in the Regression Model

Part C: How Do You Address Problems Related to Multicollinearity?

Part A: What is Multicollinearity?

Multicollinearity occurs when there is a sort of conflict (in the form of a strong correlation) between two or more predictor variables in a regression model. When these predictor variables evidence this conflict due to this high correlation, the true relationship between each predictor and the dependent variable is often misrepresented. Specifically, when a high correlation (multicollinearity) between predictor variables exists, the regression model often indicates that neither predictor is significantly related to the dependent variable, when in fact both are. *Multicollinearity* can easily weaken the power of the regression model to indicate when true findings exist between the predictor variables and dependent variable.

Conceptually, *multicollinearity is the type of problem that might occur when two usually fun party guests have a conflict at your party because each is wearing the same red bowtie.* Specifically, let's say you invite two guests to your party, we'll call them Guest A and Guest B. Each is so much fun that you can predict that your party will be great when either of them attends the function. In other words, the presence of either Guest A or Guest B predicts a great party.

However, at this one particular party, Guest A and B unexpectedly wear the same style red bowtie (see **Figure 4.43**) and become upset with one another. In fact, a conflict ensues between the two guests and as a result nobody at the party is having a good time. Now, the relationship where Guest A and Guest B would have each significantly predicted a great party has been changed due to the conflict between them. In other words, each guest individually would have predicted a great party. However, when both were present, the conflict between them resulted in neither of them predicting a great party. The example of how the conflict between Guests A and B deteriorate their otherwise significant predictive relationship of a great party can be generalized to illustrate how multicollinearity operates in a regression model.

Figure 4.43   The conflict between guest A and B

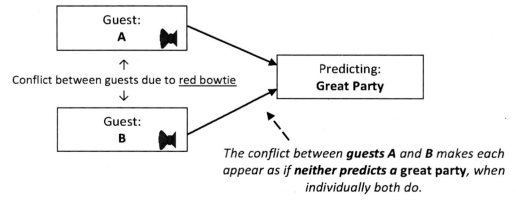

*The conflict between **guests A** and **B** makes each appear as if **neither predicts a** great party, when individually both do.*

In the **Figure 4.44**, you will see Guests A and B have been replaced with predictor variables *Age* and *Years of Education*, as well as the predicted factor replaced with our dependent variable *Happiness*. Here you see a similar relationship to the party scenario, where the conflict between the predictor variables, which is due to a high correlation (not a red bowtie), changes the relationship between each predictor variable and dependent variable being predicted.

Specifically, the predictor variable *Age* might be significantly associated with (predictive of) the dependent variable *Happiness*, but the conflict (high correlation) between the predictor variables *Age* and *Years of Education*, now makes that relationship appear not significant. The same challenge might also exist between the predictor variable *Years of Education* and the dependent variable *Happiness* due to the predictor variable conflict.

Figure 4.44   Multicollinearity between *Age* and *Years of Education*

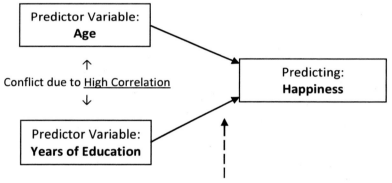

*The conflict (Multicollinearity) between **Age** and **Years of Education** makes each appear as if **neither predicts Happiness**, when individually both do.*

Let's put this in statistical terms. A predictor variable explains the dependent variable, which is another way of saying that a predictor variable explains the *variance* in the dependent variable. The *variance* in the dependent variable refers to how scores go up and down. For example, the potential range of scores for our dependent variable *Happiness* is 1 to 5 (*Variance* in *Happiness* scores occurs when scores *vary* from 1 up to 5).

When predictor variables are highly correlated, it is as if the two predictor variables are correlated so highly that each predictor variable is *explaining the same variance* in the dependent variable. This can be referred to as multicollinearity.

Part B: How Do You Assess if Multicollinearity exists?

Statistically (1): Correlations. Of course, regression models commonly have predictor variables that are correlated with one another. The correlation between predictor variables becomes a problem when the relationships reach a certain high level of correlation. There are several methods used to determine if predictor variables are correlated too highly in a regression model. For example, many analysts produce a correlation matrix that includes all the continuous predictor variables in a study in order to identify if any two are correlated at a level of .80 or .90 (i.e., the correlation coefficient is .80 or .90). To illustrate our example, let's produce a correlation matrix between the predictor variables *Age* and *Years of Education*. To produce this correlation matrix in SPSS go to:

1) **Analyze → Correlation → Bivariate**.

2) In the dialogue box that opens, use the center arrow to move the variables to be correlated (e.g., *Years of Education*, *Age*) from the left hand column into the right hand column.

3) Under the term **Correlation Coefficients**, be sure the box is checked for **Pearson** (we are conducting a Pearson's correlation) as in **Figure 4.45**.

4) Click the **OK** or **Paste** button.

Figure 4.45   Using the bivariate correlation to examine multicollinearity

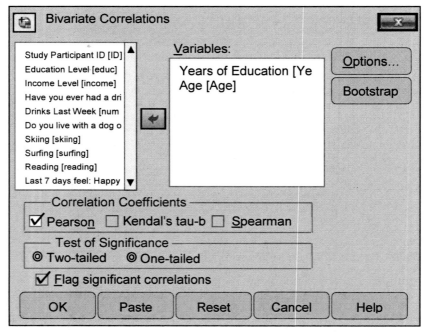

SPSS Output 4.12 presents the statistical output for Pearson's correlation test. As mentioned, the output produced is known as a correlation matrix. Within a correlation matrix there are two intersecting cells for each variable combination where each cell presents the same data. Thereby, we only need to look at one of the cells. For example, let's observe the cell which begins with the term *Years of Education*. To the right of the term *Years of Education*, you will see the term Pearson Correlation, then in the far column on this line you will see the value of the correlation coefficient (the correlation coefficient here is called the Pearson's *r*), which is .295. Thus, the correlation coefficient between these two predictors is below .80 or .90 (i.e., the levels that would indicate cause for concern, see Allison, 1999, pp. 41-42), so we can assume that these predictors will not present problems related to multicollinearity when used together in a regression model.

Please note that next to the correlation coefficient are two asterisks, which indicates that that the predictors *Years of Education* and *Age* are correlated at a statistically significant level of .01. This illustrates what was stated earlier in that the predictors can be correlated, even significantly correlated, and still not present problems related to multicollinearity. The relationship(s) between predictor(s) only becomes problematic when the correlation coefficient reaches the .80 or .90 level. In other words, when the variables are correlated at the .80 or .90 level, the red bowtie conflict may exist.

SPSS Output 4.12

Correlations

A correlation coefficient >.90 suggest *Multicollinearity*

Correlations

		Years of Education	Age
Years of Education	Pearson Correlation	1	.295**
	Sig.(2-tailed)		.003
	N		100
Age	Pearson Correlation	.295**	1
	Sig.(2-tailed)	.003	
	N	100	

**p<.01

Statistically (2): Collinearity Diagnostics in the Regression Model. SPSS has several collinearity diagnostics that can indicate if problems related to multicollinearity are present within a regression model. Two widely used estimates of multicollinearity are the variance inflation factor (VIF) and the tolerance statistic. These statistics are computed in the actual regression model. The VIF indicates if a predictor has a strong correlation with other predictors in the regression model. There is no universally accepted VIF value that indicates concern. However, a **VIF** value **larger than 10.0** among predictor variables indicates a problem (Bowerman & O'Connell, 1990; Myers, 1990). Furthermore, if you average out the VIF values of all predictor variables in a regression model and the value is substantially greater than '1,' problems related to multicollinearity might be present (Myers, 1990). A tolerance statistic (the reciprocal of the VIF) **less than .20** indicates a cause for concern and a value **less than .10** indicate a serious problem (Menard, 1995).

More rigorous cutoff points for these values, such as no **VIF** values in a regression model **greater than 2.5 or 3.0** are also regularly recommended (Allison, 1999). As mentioned earlier, SPSS produces these collinearity diagnostics within the actual regression model. Therefore, in order to illustrate how we can check if VIF and tolerance values are acceptable, we will now conduct a multiple linear regression model with *Happiness* modeled as the dependent variable and the variables *Years of Education* and *Age* entered as predictors. To conduct this procedure in SPSS go to:

1) **Analyze→ Regression→ Linear**.

2) Enter the dependent variable (*Happiness*) in the box labeled **Dependent** and the predictors in the box labeled **Independent(s)** as presented in **Figure 4.46**.

Figure 4.46   The dialogue box for linear regression

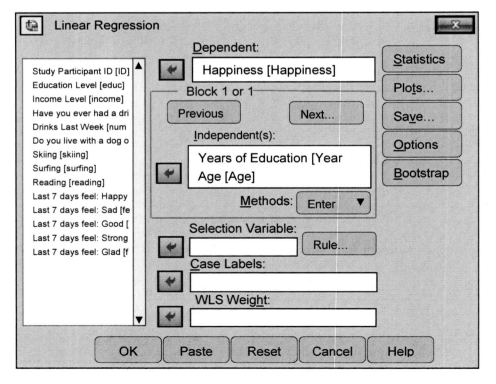

3) In the upper right corner, click the button labeled **Statistics**.

4) Check the **Collinearity Diagnostics** box as presented in **Figure 4.47**.

5) Click **Continue**.

6) Click **OK** or **Paste**.

Figure 4.47 Activate the **collinearity diagnostics** in the linear regression

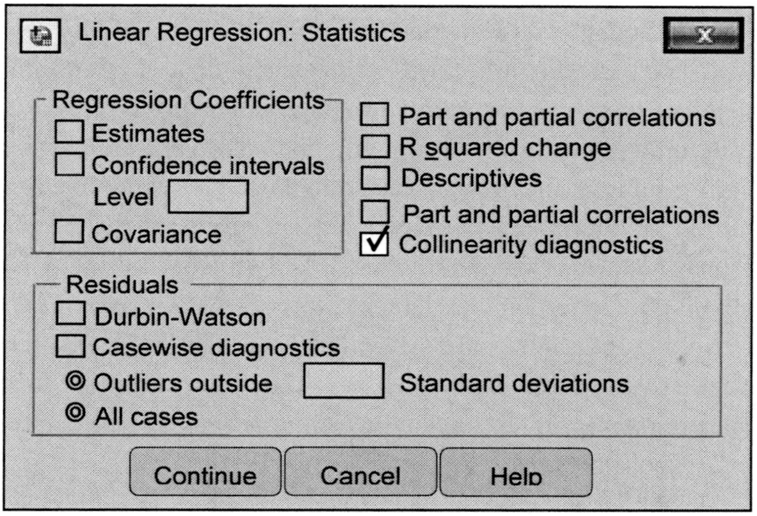

SPSS Output 4.13 presents the final box produced within the statistical output for the multiple linear regression analysis. Notice at the final column is labeled **Collinearity Statistics** has two subcolumns (Tolerance and VIF). Recall, earlier we stated that **VIF**

SPSS Output 4.13

Coefficients[a]

Model	Unstanderd. Coeffic.		Standard. Coeffic.			Collinearity Statistics	
	B	SE	Beta	t	Sig.	Tolerance	VIF
(Constant)	1.45	.31		4.7	.00		
Age	.00	.01	.001	.03	.98	.996	1.004
Years of Educ.	6.4	.00	.884	18.6	.00	.996	1.004

Tolerance Scores above .20 suggests no significant *multicollinearity*

VIF Scores under 2.50, suggests no significant *Multicollinearity*

values **greater than 2.5** and **tolerance** values **less than .20** are a cause for concern. We can see within these subcolumns that regarding our two predictor variables, the value for tolerance is above .20 (Tolerance=.996) and the value for VIF is less than 2.5. Therefore, these criteria indicate that regarding the predictor variables *Age* and *Years of Education*, multicollinearity does not present a significant problem.

Part C: How Do You to Address Problems Related to Multicollinearity?

There are several methods to address unacceptably high levels of multicollinearity in a regression model. The easiest solution is to remove the most intercorrelated variable(s) from analysis. However, this is commonly not practical or preferred. At times, **centering** is used to transform the offending independent variables. To center a variable you simply subtract the mean score from each case using the **Transform→ Compute Variable** function in SPSS. This method *centers* the mean score within the distribution at '0', which often results in considerably lower multicollinearity. At times, an analyst might combine predictor variables if this seems like a logical choice. For example, if two predictor variables are subscales from the same measurement instruments and are highly correlated, you might consider combining them into the total score of the measurement tool. At times increasing the sample size can reduce levels of multicollinearity (if possible).

4.4.2.3 Homoscedasticity

Homoscedasticity is an important assumption related to an equivalent of variance. This assumption is another important consideration that must be made before a researcher can have confidence in the results of a parametric statistical procedure. In this section we will review how to assess homoscedasticity in the following three parts:

Part A: What is a Homoscedasticity?

Part B: How Do You Assess if Homoscedasticity Exists?

Descriptively: Boxplot Graphical Display

Statistically: Levene's Test of Homogeneity

Part C: How Do You Address Problems Related to Homoscedasticity?

Part A: What is a Homoscedasticity?

The assumption of homoscedasticity in regression assumes that residuals have similar variances at each level of a predictor variable. A *residual* or error represents unexplained variance after conducting a regression model. Homoscedasticity may be referred to as *homogeneity of variance*. The root *Homo* by definition means *the same*.

Variance can be described as how scores vary from high to low, such as *Happiness* scores varying from 1-5. Thus, *homogeneity of variance* suggests that scores *vary* (variance) in *the same* (homo) way. The lack of homoscedasticity is referred to as heteroscedasticity, which can weaken a study and lead to serious distortion of statistical findings (Berry & Feldman, 1985; Tabachnick & Fidell, 1996).

Homoscedasticity suggests equal levels of variability regarding a continuous dependent variable, across levels a predictor variable(s), which may be either continuous or categorical. In this respect, you might think of our check of homoscedasticity as sort of *cross-referencing* of variables as a means of assessing if an equivalence of variance exists. For example, in terms of assessing homoscedasticity as an assumption for our regression analysis, we might examine if residuals for the dependent variable *Happiness* are similar across the range of a predictor such as *Years of Education*. **Figure 4.48** presents a homoscedastic (Homogeneity of Variance) relationship between the variables *Years of Education* and *Happiness*. You can see within this figure that at each level of the predictor variable *Years of Education* there is a rather equal variance of residuals derivative of the dependent variable *Happiness*. Thus, homoscedasticity is indicated.

Figure 4.48   When *Homoscedasticity* Occurs

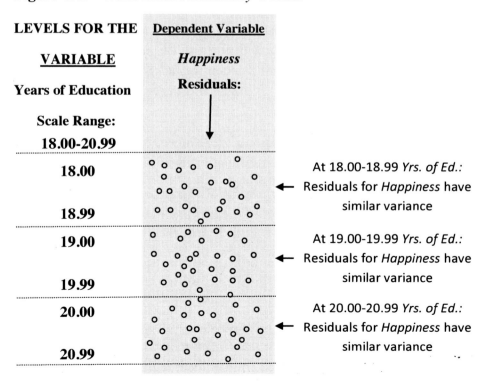

As mentioned earlier, when there is not a similar level of variance in dependent variable scores at different levels of a predictor variable, we call this heteroscedasticity (the opposite of homoscedasticity). **Figure 4.49** illustrates a distribution when *heteroscedasticity* occurs (Non-Homogeneity of Variance). Within the figure you can see that there is a different level of variance regarding the residuals derivative of *Happiness* (dependent variable) within each level of the predictor variable *Years of Education*. Specifically, within the portion of the *Years of Education* variable from 18.00-18.99 there is a rather healthy variance of residuals. However, at the second level of measurement for *Years of Education* (19.99-19.99) the variance of residuals shrinks. At the third level of measurement (20.00-20.99) of *Years of Education* the variance of residuals shrinks further, creating a funnel shaped distribution and reflects heteroscedasticity.

Figure 4.49   When *Heteroscedasticity* Occurs

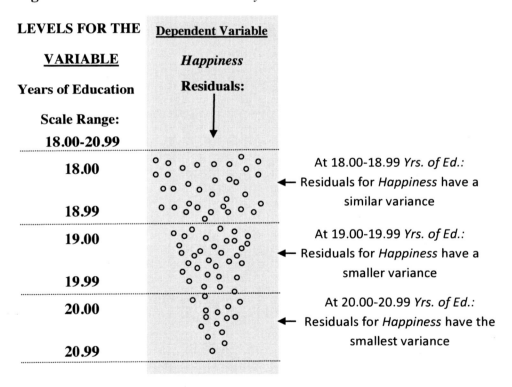

Part B: How Do You Assess if Homoscedasticity exists?

Descriptively: Boxplot Graphical Display. Homoscedasticity is typically examined via a boxplot graphical display of the *Regression Standardized Residual* by the *Regression Standardized Predicted Value*, which is produced in the linear regression function.

Let's use our example of relating the variance in dependent variable *Happiness* score to the various levels of the predictor variable *Years of Education*.

Specifically, in SPSS go to:

1) **Analyze→ Regression→ Linear.**

2) Within the dialogue box that opens (see **Figure 4.46**) enter the dependent variable *Happiness* in the box in the top center labeled **Dependent**.

3) Enter the predictor variable *Years of Education* into the box below labeled **Independent(s)**.

4) Click the button in the upper right hand corner marked **Plots**.

5) In the dialogue box that opens (see **Figure 4.50**) in the left hand column, move the term *ZRESID in the box under Y.

6) Then move the term *ZPRED into the box under the X.

7) Click the button marked **Continue**.

8) Click either **OK** or **Paste**.

Figure 4.50   Activating the **ZPRED** by **ZRESID** plot in linear regression

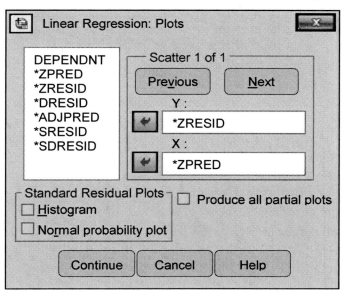

Figure 4.51 presents the *Regression Standardized Residual* by the *Regression Standardized Predicted Value* plot that would be included within the linear regression statistical output. Within the plot, the points are randomly and evenly dispersed around the center. This pattern suggests that the assumption of homoscedasticity has been met

regarding the two variables modeled in the plot. Notice, the plot in **Figure 4.51** reflects a full dispersion of values across all levels making a full square shape similar to **Figure 4.48**. Please note the **Figure 4.52**, which presents a scatterplot indicating *Heteroscedasticity* as points are biased, which here creates a funnel shaped distribution. This plot indicates a biased variance where a lower degree of variance appears on the left as scores are close together, but points are more spread out to the right.

Figure 4.51   ZPRED by ZRESID Scatterplot Indicating *Homoscedasticity*

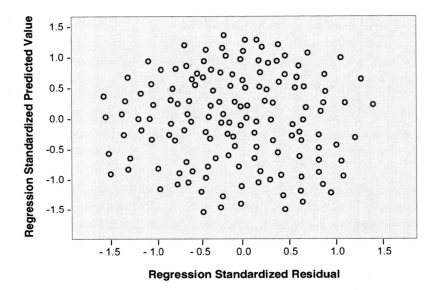

Figure 4.52   ZPRED by ZRESID Scatterplot Indicating *Heteroscedasticity*

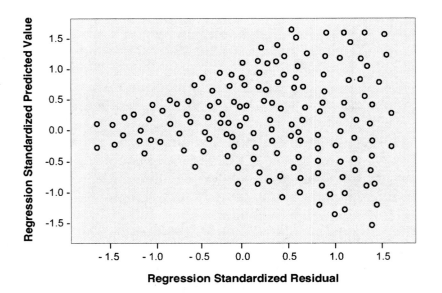

Specifically, the plot suggests the variance of dependent variable *Happiness* is different at various levels of the predictor variable *Years of Education*. Subsequently, as points are not randomly and evenly dispersed throughout the plot, we cannot assume that the assumption of homoscedasticity has been met.

Statistically: Levene's Test of Homogeneity

Homoscedasticity can also be examined through the Levene's test of homogeneity available in the one-way ANOVA procedure. To conduct this test in SPSS go to:

1) **Analyze→ Compare Means→ One-Way ANOVA.**

2) Move the dependent variable *Happiness* from the left hand column into the top box in the right hand column labeled **Dependent List**.

3) Next, use the lower center arrow to move the independent variable *Years of Education* from the left hand column into the lower box in the right hand column labeled **Factor**, as presented in **Figure 4.53**.

4) Click **Options**.

Figure 4.53   One-Way ANOVA: *Happiness* by *Years of Education*

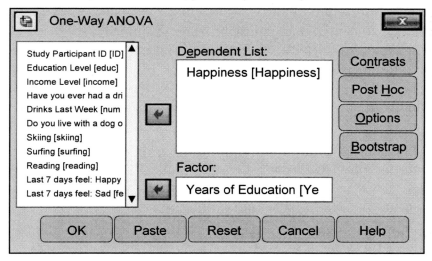

5) In the dialogue box that opens (see **Figure 4.54**), click the box for **Homogeneity of variance test**.

6) Click **Continue**.

7) Click **OK** or **Paste**.

Figure 4.54 Activate the **homogeneity of variance test** in the One-Way ANOVA

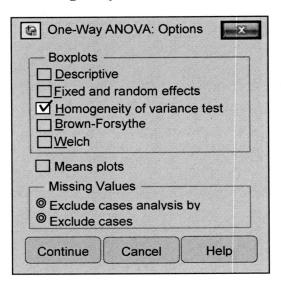

SPSS Output 4.14 presents the results of the **Tests of Homogeneity of Variances** within the statistical output produced by the One-Way ANOVA. The final column labeled **Sig.**, presents the level of probability for the **Tests of Homogeneity of Variances**. If the level of probability is .05 or less, the test indicates that the variances are significantly non-homogenous, indicating heteroscedasticity, and that the assumption of homoscedasticity has not been met. If the level of probability is greater than .05 ($p<.05$), homoscedasticity is indicated.

SPSS Output 4.14

Oneway

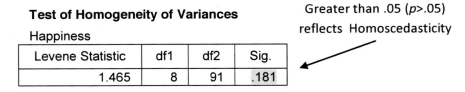

Test of Homogeneity of Variances

Happiness

Levene Statistic	df1	df2	Sig.
1.465	8	91	.181

Greater than .05 ($p>.05$) reflects Homoscedasticity

Part C: How Do You to Address Problems Related to Homoscedasticity?

When the assumption of homoscedasticity is not supported, the dependent variable can be transformed as to better approximate homoscedasticity. Recall in **Section 4.4.2.1 Normal Distribution** under the title *Option 2: Data Transformations* we described three common data transformations used to transform a variable to better approximate normality: the **log transformation**, the **square root transformation**, and the **inverse (reciprocal) transformation**. These same data transformations can be used to transform a dependent variable toward better approximating a homoscedastic distribution. To transform the dependent variable in this manner follow the same directions presented in that section. Again, you would transform the variable then re-examine the transformed version of the dependent variable to assess if the distribution better approximates homoscedasticity.

4.4.2.4 Linearity

Linearity refers to the occurrence where two variables evidence a **linear relationship**. As you might guess, multiple linear regression analysis (the multivariate procedure used in sample study one) is a statistical test that examines a linear relationship between variables. Subsequently, if multiple linear regression analysis is applied to examine the relationship between two variables that do not have a linear relationship, then the wrong statistical procedure might be in use. Thus, before we apply multiple linear regression analysis in sample study one, it is imperative to assess if the assumption of linearity between predictor variables and the dependent variable has been met. In this section we will review how to assess linearity in the following three parts:

Part A: What is Linearity?

Part B: How Do You Assess if Linearity Exists?

Descriptively 1: Boxplot Graphical Display

Descriptively 2: Simple Scatterplot

Part C: How Do You Address Problems Related to Linearity?

Part A: What is Linearity?

Essentially, a linear relationship is an association where, as scores increase in one variable (e.g., the predictor variable), the scores in the second variable (e.g., the dependent variable) correspond by either increasing or decreasing in a constant manner for the entire range of scores of the scale. For example, **Figure 4.55** presents a linear relationship between *Income Level* (Y axis) and the *Happiness* (X axis). You will notice that as *Income Level* increases, *Happiness* scores increase perfectly. For example, for Case 1, there is an intersection between the lowest possible score on the *Happiness* scale (1) and the lowest possible value for the variable *Income Level* ($100k). After that there is constant intersection of scores along the two continuums until the highest income level ($500k) intersects with the highest *Happiness* score (5). This is known as a *positive relationship*, where an increase of scores on one scale is associated with an increase of scores on the other scale.

Figure 4.55   A Perfect *Positive* Linear Relationship (Correlation)

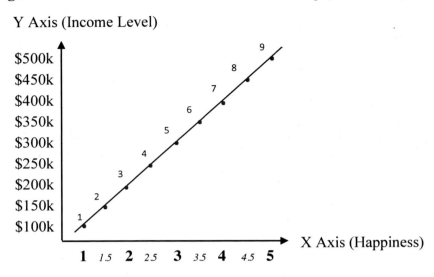

Figure 4.56 also presents a linear relationship between *Income Level* (Y axis) and *Happiness* scores (X axis). You will notice that as *Income Level* increases, *Happiness* scores decrease perfectly. For example, for Case 1, there is an intersection between the lowest possible score on the *Happiness* scale (1) and the highest value reflecting *Income Level* ($500k). After that there is a constant intersection of scores along the two continuums until the lowest value reflecting *Income Level* ($100k) intersects with the highest *Happiness* score (5). This is a *negative relationship*, where an increase of scores on one scale is associated with a decrease of scores on the other scale.

Figure 4.56   A Perfect *Negative* Linear Relationship (Correlation)

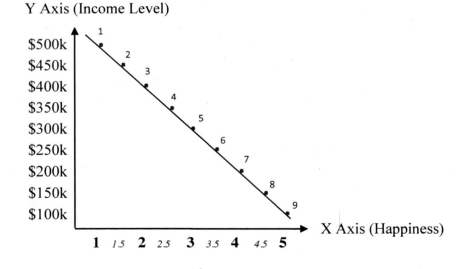

Other types of relationships might exist between variables, such as a *curvilinear* relationship, which begins in a pattern reflecting a *positive* relationship, but changes into a pattern reflecting a *negative* relationship. **Figure 4.57** presents a *curvilinear* relationship where cases 1 to 5 reflect a *positive* linear relationship (as *Happiness* scores increase as *Income Level* increases), but after $300k on the Y axis, *Happiness* scores begin to decrease.

Figure 4.57   A *Curvilinear* Relationship (correlation)

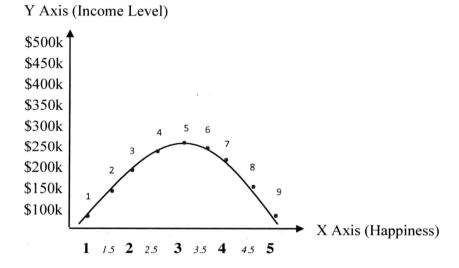

Part B: How Do You Assess if Linearity exists?

Descriptively (1): Boxplot Graphical Display. When assessing linearity, you can use the *Regression Standardized Residual* by the *Regression Standardized Predicted Value* plot. This is the same plot used in **Section 4.4.2.3** to assess homoscedasticity. Therefore, for instructions on producing this plot refer to **Section 4.4.2.3 Homoscedasticity** under the subheading **Part B: How Do You Assess if Homoscedasticity exists**. Within the statistical output the *Regression Standardized Residual* by the *Regression Standardized Predicted Value* plot will again appear.

When the points are randomly and evenly dispersed throughout the plot as in **Figure 4.58** linearity is indicated. **Figure 4.59** presents a scatterplot indicating a *nonlinear* relationship, where the points reflect a *curved* formation. Subsequently, as points are not randomly and evenly dispersed throughout the plot, we cannot assume that the assumption of linearity has been met and we might speculate that the relationship between variables is nonlinear.

Figure 4.58   ZPRED by ZRESID Scatterplot Indicating *Linearity*

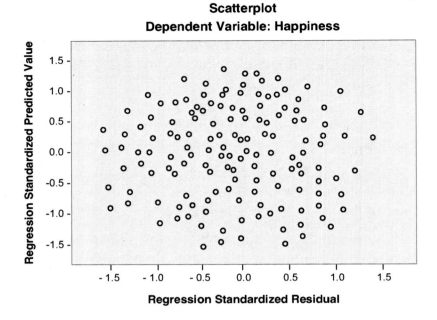

Figure 4.59   ZPRED by ZRESID Scatterplot Indicating *Nonlinearity*

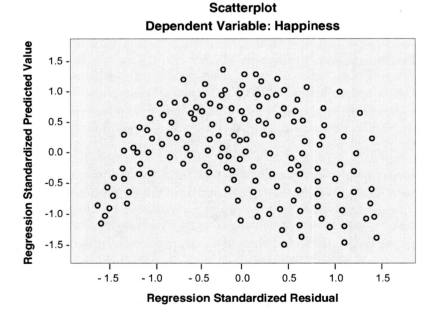

Simple Scatterplot. There are many useful plots used to assess linearity. For example, a scatterplot is often a useful tool to assess if a relationship between variables is linear. One method of producing a scatterplot in SPSS is to follow the path:

1) **Graphs→ Legacy Dialogs→ Scatter/Dot.**

2) In the dialogue box that opens (see **Figure 4.60**) click the box labeled **Simple Scatter**.

3) Click **Define**.

4) In the dialogue box that opens (see **Figure 4.61**) enter the first continuous variable in the X Axis. For example, we have entered the dependent variable *Happiness* in this cell.

5) Enter the second continuous variable in the Y Axis. We have entered the covariate predictor variable *Income Level* in this cell.

6) Click **OK** or **Paste**

Figure 4.60   Selecting the **Simple Scatter** function

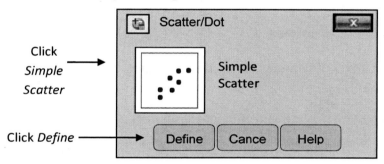

Figure 4.61   Entering variables into the scatterplot

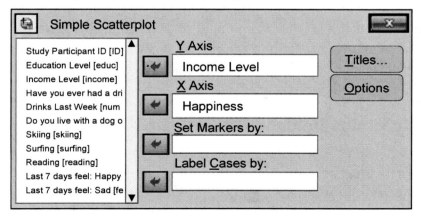

The statistical output will include a graph similar to **Figure 4.62**. This figure presents a scatterplot that describes the relationship between *Happiness* scores and *Income Level*. Within this figure a linear relationship is indicated by the cigar shaped pattern of scores formed, which reflects as *Income Level* increases, *Happiness* scores increase in a uniform manner. This pattern is similar to the **Figure 4.55** that presents a perfect *positive* linear relationship (correlation). Thus, from this scatterplot pattern we can speculate that a linear relationship exists between these two continuous variables.

Figure 4.62   A scatterplot presenting a *positive* linear relationship

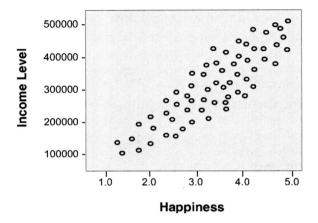

Part C: How Do You Address Problems Related to Linearity?

When a relationship between two continuous variables is not linear, a common solution is to apply one of the data transformation techniques discussed earlier (logarithmic, square root, inverse) to one or both variables to achieve a relationship that is linear. Recall, these transformations produce a new variable that is mathematically equivalent to the original variable, but expressed in different units of measurement that may better approximate a linear relationship. If data transformations do not work, when appropriate, one or both of the variables might be dichotomized or made into another categorical structure to better accommodate analysis. Furthermore, other statistical procedures that examine nonlinear relationships, such as curvilinear regression, should be considered if this relationship pattern between variables is indicated.

4.4.2.5   No Undue Influence of Outlier Scores

The earlier section **4.4.2.1 Normal Distribution**, not only defined an outlier score, but also described the need to assess the influence of an outlier score in making a distribution of scores non-normal. This section will continue with this idea and describe not only the need to assess if outlier scores are impacting normality, but the results of the statistical analysis. It is essential to assess if outlier scores are influencing the results of statistical analysis because in certain situations, just one outlier score can misrepresent the data. Specifically, it is entirely possible that just one outlier score can result in a statistical test indicating that a significant relationship exists between variables, when a true relationship does not exist.

When one or several outlier scores are driving the results of a statistical analysis, it is said that the outlier scores have an *undue influence* on the study results. Before a researcher can have confidence in the results of parametric statistical tests, it is essential to assess if outlier scores are evidencing an undue influence on study results. Conveniently, the process of assessing the impact of outlier scores as described in the section **4.4.2.1 Normal Distribution** is the same process we will use to assess if outlier scores have an undue influence on study results. Please refer to that section for a full discussion how to define an outlier score, how to identify an outlier score, as well as test the influence of an outlier score in the section *Option 1: Assessing the Impact of Outlier Scores*.

In this section, we will discuss the specific steps needed to test for the influence of outlier scores on study results, while making reference to the applicable portions of the section *Option 1: Assessing the Impact of Outlier Scores* (within **Section 4.4.2.1**) that need to be followed rather than repeating the directions here. Specifically, using the sample study one with the dependent variable *Happiness* as an example, to assess the influence of outlier scores on study results you must take the following three steps:

1) Conduct the data analysis (e.g., the full multiple linear regression model) using the full distribution of scores, i.e., with the **outlier scores included** within the distribution of scores.

2) Conduct the data analysis (e.g., the full multiple linear regression model) with the **outlier scores excluded** from the analysis, i.e., without the outlier scores within the distribution of scores.

3) Note if the results of the data analysis (e.g., the two multiple linear regression models) **differ significantly ($p<.05$) when the outlier scores are included or excluded** from the analysis.

Of course, as we have seen in the piece titled *Option 1: Assessing the Impact of Outlier Scores* within the section **4.4.2.1 Normal Distribution**, completing this process involves four smaller steps. These steps would be amended slightly when applying them to assessing the *impact of outlier scores on study results* relative to assessing the impact of outlier scores on normality. Specifically, to assess the impact of outlier scores on study results you must:

1) **Identify the Outlier Scores: STEP 1** of the section *Option 1: Assessing the Impact of Outlier Scores*, instructs on the need to **Identify the Outlier Score(s)** either through *The Boxplot* or *The Raw Mean and Standard Deviation Values*. This step would be conducted in exactly the same manner within this section.

2) **Assess the Impact of the Outlier Scores: STEP 2** within the section *Option 1: Assessing the Impact of Outlier Scores*, instructs on the need to **Test the Distribution of Scores for Normality Without the Outlier Score(s)** via three tasks. This instruction would be followed in the same original three task format in this section with the exception that the impact of outlier scores would be examined in relation to study results and not normality. Specifically, just as in the section *Option 1: Assessing the Impact of Outlier Scores*, we would first **create a filter variable** to identify the outlier scores (Task 1). Second, you would use the **select cases** function in SPSS to exclude the outlier scores from analysis (Task 2). However, in the third task instead of *Testing the Distribution of Scores for Normality with the Outlier Scores Excluded* you would *Examine the Results of the Data Analysis with the Outlier Scores Excluded.*

Using the two multiple linear regression models from the original section as examples, recall as presented in the **SPSS Output 4.10**, the first regression analysis indicated statistically significant relationships ($p<.05$) between each predictor variable and the dependent variable *Happiness* when the four outlier *Happiness* scores were **excluded** (where 96 scores are in the analysis). This is indicated by all the values in the final column labeled **Sig** being below the value .05.

SPSS Output 4.10 (presented for a second time)

Coefficients[a]

Model	Unstandardized Coefficients		Standardized Coefficients	t	Sig.
	B	Std. Error	Beta (β)		
Years of Education	.06	.02	.212	2.6	.011
Income Level	.01	.00	.481	5.8	.000
Live with a dog or a cat?	-.34	.11	-.242	-2.9	.004

All relationships are significant ($p<.05$) with **96** ↗
study participants (outlier scores are excluded)

The second multiple linear regression model, presented in **SPSS Output 4.11**, presented this same analysis using the full sample of *Happiness* scores that includes the four outlier scores (where we have all 100 scores in analysis). Within this output, the results of this multiple linear regression analysis also indicated each predictor evidenced a statistically significant ($p<.05$) relationship with the dependent variable *Happiness*. Again, this is indicated by all the values in the final column to the right labeled **Sig** being below the value .05.

SPSS Output 4.11 (presented for a second time)

Model	Coefficients				
	Unstandardized Coefficients		Standardized Coefficients	t	Sig.
	B	Std. Error	Beta (β)		
Years of Education	.05	.02	.18	2.3	.027
Income Level	.01	.00	.48	6.0	.000
Live with a dog or a cat?	-.42	.12	-.28	-3.5	.001

All relationships are significant ($p<.05$) with **100** 🡕
study participants (outlier scores are included)

3) **Assess the Impact of Removing the Outlier Score(s): STEP 3** of the section *Option 1: Assessing the Impact of Outlier Scores* then instructs to **Assess if Removing the Outlier Score(s) Supported a Normal Distribution of Scores.** However, in this section we would compare the regression models to examine **if the results of the statistical analyses differ between the regression models where the outlier scores were included and excluded.** For example, across both regression models we see the same statistically significant ($p<.05$) relationships between the predictors and the dependent variable. In other words, the results of the analysis do not change when the outlier scores are included or excluded from analysis.

4) **Find Evidence of Robustness: STEP 4** of the section *Option 1: Assessing the Impact of Outlier Scores* instructs on the need to **Find Evidence of Robustness in the Parametric Testing (e.g., multiple linear regression) by Removing and Reinserting the Outlier Scores from the Analysis.** This process would be repeated in this section.

Recall in the original section that *robustness* is indicated when the results of the parametric statistical test are not significantly impacted by violation of test assumptions, such as outlier scores. Thus, as the findings of this analysis indicated that the outlier scores did not significantly impact study results, you might describe that the multiple linear regression analysis was quite *robust*. Specifically, these extreme values *did not seem to evidence an undue influence on study results*. Again we can make this statement because the outlier scores did not change the results (e.g., statistical significance, directionality, and effect size) of inferential analysis in terms of how the predictors were related to the dependent variable within the final multivariate model.

However, if the two multiple linear regression analyses indicated the relationships between the predictor and dependent variables differed significantly (in terms of statistical significance, etc.) when outliers were included and excluded, we would need another approach. For example, in this case you might decide to exclude the outliers from analysis and report that you did so. You might do this as the analysis would be indicating that the outlier scores are evidencing an undue effect upon the study results, which is a violation of the test assumption we are describing. Alternately, you might perform a data transformation technique to reduce the influence of outlier scores as described within the section **4.4.2.1 Normal Distribution** under the heading *Option 2: Data Transformations*.

4.4.2.6   Other Test Assumptions

There are several other assumptions associated with regression that should be considered (Berry, 1993), including:

1) **Predictors are continuous or categorical with a two category response**. In a regression model, the predictors can be continuous or categorical, but the categorical variable must be dichotomous (two categories). You will see later in **Section 4.7.1**, when needed this is accomplished through *dummy-coding*.

2) **Predictors have variation in value**. All predictors must have some variation in value, which means the values within a variable are not constant. This means the responses within a predictor variable are not all the same.

3) **Predictor variables are not correlated with any variables not included in the regression analysis that influence the dependent variable**. The error in the regression model increases when there are variables in the dataset that are not included in the regression model, but are significantly correlated with the variables that are included in the regression model. Therefore, it is important that all predictor variables that are significantly related to the dependent variable are represented in the regression model.

4) **Independence of observations**: Independence of observations is the assumption that there is no relationship between the study participants in different groups. For example, regarding the variable *Do You Live with a Dog or Cat?* each study participant is separate and is not included within more than one group (not included in the *Dog* and *Cat* categories). This is a rather conceptual issue related to the study design. In sample study one and two presented in this text, all observations within groups are independent in this manner.

5) **Independence of errors**. The residual terms should be independent (no serial correlation). This can be tested via the **Durban-Watson test**, which is available in the linear regression function in SPSS at:

   a. **Analyze→ Regression→ Linear**.

   b. In the dialogue box that opens (see **Figure 4.46**) enter the predictor variables (*Education Level, Income level, Do You Live with a Dog or a Cat?*) and dependent variable (*Happiness*).

   c. In the upper right hand corner click the button labeled **Statistics**.

   d. In that dialogue box that opens, check the box for **Durbin-Watson** as presented in **Figure 4.63**.

   e. Click **Continue**.

   f. Click **OK** or **Paste**.

Figure 4.63   Activating the **Durban-Watson** test in the linear regression

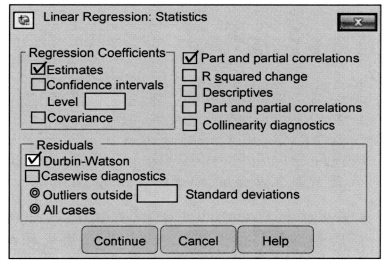

SPSS Output 4.15 presents the statistical output for this procedure where the value for the **Durbin-Watson test** is presented in the final column of the **Model Summary** box. The **Durbin-Watson test** statistic ranges from 0 to 4. As a general rule, the residuals may be considered uncorrelated if the statistic is approximately 2 (an acceptable range is 1.50 - 2.50). If the Durbin-Watson statistic is not in the acceptable range, you need to report this as a reason to regard findings with caution as a violation of the regression assumptions exists.

SPSS Output 4.15

Model Summary

Model	R	R Square	Adjusted R Square	Std. Error of the Estimate	Durbin-Watson
1	.905ᵃ	.820	.812	.32749	.878

If the Durbin-Watson Statistic Is Between 1.50 to
2.50 **Independence of Errors** Can Be Assumed

Summary. This section may have reflected that checks of test assumptions can be quite involved. In fact, preparing the data through checks of test assumptions often requires more time and attention than the actual inferential analysis. Thus, these checks of test assumptions very often represent the most involved work associated with conducting a data analysis study. We also mentioned that a violation in test assumptions weaken or distort the results of statistical analysis quite easily, so checks of test assumptions are vital if findings are to be regarded with credibility.

4.4.3 Missing Data

If there is one subject in data analysis that is both extensive and complex it is how to treat missing data. *Missing data generally refers to the presence of a missing data value within a study variable(s) for a particular case(s)*. In other words, a study participant most likely did not provide a response to a survey item, and then when the data from the study survey are entered into the SPSS database, the non-response becomes a blank cell in the database. The big question becomes, what do you do with that empty cell?

In SPSS, there is a *listwise deletion* function that excludes cases with missing data from the analysis. The missing data cell(s) becomes like a scarlet letter banishing the cases from the analysis. Essentially through this method, SPSS is ignoring the cases with missing data. Instead of treating those cases, the software program deletes them from the analysis. Although this may be the simplest solution, it is not the most sophisticated. You will see in the text below that the methods to account for missing data have become very sophisticated to the point of specialized software and intricate methods. You may wonder when you might need one of these very sophisticated methods. Toward that, one should consider that overall **the greater the challenge presented by the missing data, the more sophisticated the method needed to address the challenge.**

Fortunately, only a small number of datasets have massive challenges related to missing data. For example, I once worked on an intervention study that examined changes in pretest scores and posttest scores reflecting changes in behavioral problems for a sample of 300 youth. The challenge related to missing data was that approximately half of the youth in the sample (i.e., 150 of 300) had missing posttest scores. However, I still needed to produce findings to indicate if the intervention was effective in reducing behavioral problems for all 300 youth. Meeting this challenge required the latest technology in addressing missing data, specifically we used the *Maximum Likelihood* method mentioned later in this section.

The scope of this book is to address the typical challenges related to missing data, which often involve a few survey items being randomly skipped among a small proportion of study participants. While the cutting edge methods are mentioned in this section, our text and analysis will focus on more practical solutions to address more typical challenges that do not require specialized software. As we stated earlier, this text is meant to be foundational, so describing practical means of addressing missing data are essential.

However, before you can treat missing data challenges, you must first identify the level of challenge presented by the missing data within your dataset. This brings us to the question of **how can you indicate the level of challenge presented by the missing data?** Perhaps the greatest indicator of the level of challenge presented by missing data concerns the question of how much data are missing, as well as what are the patterns of missing data. After these elements are assessed, you must then select an appropriate statistical procedure to address these challenges. We will review this process in this section, including how to **define missing data in a study (Section 4.4.3.1)**, assess the **amount of missing data (Section 4.4.3.2)**, assess the **pattern of the missing data (Section 4.4.3.3)**, and make a selection regarding the **treatment of missing data (Section 4.4.3.4)**.

4.4.3.1 Defining Missing Data in a Study

Before missing data can be assessed in a study, it must first be defined. Specifically, the degree of missing data within a quantitative study may be perceived in several different ways. For example, a quantitative study might include 100 variables overall. This might suggest that the proportion of missing data values would simply be the number of items without a valid response, within that total of 100 variables for each study participant. For example, if a study participant did not provide a valid response to ten of the 100 total variables, then the rate of missing data for that study participant is 10% (100/10).

However, defining the amount of missing data within a study as a proportion of the overall items without valid responses is often impractical for a number of reasons. For example, not all study variables are created equal (they don't even have a constitution suggesting so!). To illustrate this, imagine a study attempting to explain the dependent variable *Smoking Cigarettes* (Yes or No). Also, imagine within this study data on demographic characteristics such as *Geographic Location* (Urban, Suburban, or Rural) are collected as covariate variables.

If we defined missing data as the proportion of overall items without valid responses, then we would be giving equal weight for a missing valid response for the demographic covariate variable *Geographic Location*, relative to the dependent variable *Smoking Cigarettes*. This is not desirable, as the implication of missing a valid response for the dependent variable *Smoking Cigarettes* is more serious than the demographic covariate variable *Geographic Location*, as *Smoking Cigarettes* is the focus of the study and *Geographic Location* is not.

Therefore, in many studies, missing data are often defined as the number of valid responses missing from one key variable or a set of key variables. Of course, the number of valid responses among all study variables needs to be reported and accounted for, *but the central definition of missing data for the study might focus upon the valid responses missing for the key variable*(s).

For example, the focus of sample study one is explaining the dependent variable *Happiness*. The dependent variable *Happiness* is measured as a composite score, which is the product of combining five individual items. Therefore, if we select the dependent variable *Happiness* as the key variable to define missing data for sample study one, we might define missing data as **the proportion of valid responses missing among the five items** that form the composite scale measuring *Happiness*.

However, before selecting the method to define missing data for a particular study, it is important to examine the valid responses that are and are not missing for each variable within a dataset. At times, noting which valid responses are missing may play a role in how missing data is defined within a study. For example, if there are no valid items missing among the five items that compose the *Happiness* scale, but there were valid responses missing for other study items, it would not make sense to define missing data as the number of valid responses missing from the five items that form the *Happiness* composite scale. Therefore, the first step in defining missing data for a quantitative study is to **conduct a frequencies procedure for all study variables** and examine which variables are missing valid responses. This can be conducted in SPSS by going to:

1) **Analyze→ Descriptive Statistics→ Frequencies**.

2) Within the dialogue box that opens (displayed in **Figure 4.25**) move all study variables for sample study one from the left hand column into the right hand column.

3) Click **OK** or **Paste**.

SPSS Output 4.16 presents the results of the frequencies procedure, which describes how many valid and missing responses are present for each study variable. For example, within this figure, note the first row that begins with the term **Valid**, which indicates the *number of valid responses*, while the lower row labeled **Missing**, indicates the *number of missing responses* for each item. The first three study variables listed in the series of columns are *Education Level*, *Income Level*, and *Do You Live with a Dog or a Cat*. We see for each, the number of valid responses is 105 and the number of missing responses is '0,' which indicates there are no missing data points for these items. However, we see there are missing data points for the remaining five items, which form the composite score for the dependent variable *Happiness*. For example, the question asking about the level of feeling *happy* indicates only 100 valid responses and five missing values. Additionally, for the questions asking about feeling *sad*, there are 95 valid responses and 10 missing values.

SPSS Output 4.16

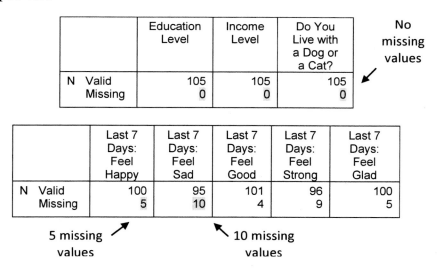

Figure 4.64 presents what a missing item might look like on a completed study survey. For example, notice that in the figure the study participant did not provide a valid response (values 1 thru 5 are not circled) for the last question asking about feeling *glad* over the past week.

Figure 4.64 Missing data point for the survey item *Glad*

In the last 7 days I felt:					
	Strongly Disagree	**Disagree**	**Neither Agree or Disagree**	**Agree**	**Strongly Agree**
Happy	1	2	3	4	⑤
Sad	①	2	3	4	5
Good	1	2	3	④	5
Strong	1	2	3	4	⑤
Glad	1	2	3	4	5

No circled value within 1-5 for
Glad = missing data point in SPSS

The frequency analysis indicated the only missing data points are among the five items that are combined to form the composite scale measuring the dependent variable *Happiness*. Subsequently, based upon the statistical output of the frequencies procedure, the approach of defining missing data as the number of missing data values within the key variable *Happiness* seems appropriate. Therefore, regarding the assessment of missing data points within the current dataset, we can **define missing data as the proportion of missing valid responses among the five items that compose the composite scale examining *Happiness* for each study participant.**

4.4.3.2 Amount of Missing Data

Now that we have defined how we will identify missing data within the dataset, we can identify the amount of missing data. The amount of missing data within a dataset is expressed through making two essential considerations, which are the *proportion of missing data values per study participant*, as well as the *proportion of study participants that have missing data values*. Each of these dimensions, as well as how to examine them, will be discussed in Part A and Part B, respectively, in this section.

Part A: The Proportion of Missing Data Values per Study Participant

Thus far, we have defined the proportion of missing data values in the dataset, as the proportion of data points that are missing among the five items that form the composite measure of the dependent variable *Happiness*. Therefore, to compute the proportion of missing data values per study participant, we must count the number of valid responses provided for these five items for each study participant. For example, **Figure 4.65** presents the study participant **ID** followed by the five items that compose the

Happiness composite scale. For study participant **ID 1**, all five items have valid responses, reflecting 100% valid responses and 0% missing data points.

Figure 4.65 Missing data point for the SPSS variable *feelglad*

ID	feelhappy	feelsadREV	feelgood	feelstong	feelglad	
1.00	5.00	4.00	4.00	4.00	5.00	100% Response (5 of 5)
2.00	1.00	2.00	2.00	3.00		(4 of 5)
3.00	4.00	4.00	4.00	4.00	3.00	80% Response

However, study participant **ID 2** has valid responses for the first four items, but an empty cell for the final item *feelglad*. This reflects that study participant **ID 2** has valid responses for four of five (80%) items and a missing data point for one of five (20%) items. We must now compute this proportion of valid responses/missing data points for each study participant. In **Figure 4.65**, we manually counted the number of valid responses for each study participant among five separate variables, to determine the number/percentage of data points missing. However, there are several ways to count the number of valid responses for multiple variables more easily using the SPSS software.

The **Count Occurrences of Values within Cases** function can be used to produce an aggregate score of how many valid responses each study participant provided to a number of variables. For example, this function would produce a score reflecting how many of the five variables that comprise our *Happiness* scale have a valid response. To use this function, go to:

1) **Transform → Count Occurrences of Values within Cases.**

2) Within the dialogue box that opens, in the box labeled *Target Variable* enter a name for the new variable where the count of values will be reflected. For example, we have entered the variable *HappinessNumResp* to indicate this variable reflects the number (*Num*) of responses (*Resp*) for the *Happiness* scale.

3) In the box *Target Label* enter the variable description, such as *Number of Items with Responses on the Happiness Scale*.

4) Then highlight the five variables in the left hand column that comprise the *Happiness* scale and use the center arrow to move them into the right hand column marked **Numeric Variables**.

5) The dialogue box will look similar to the **Figure 4.66**.

6) Click **Define Values….**

7) In that dialogue box that opens, click the toggle button marked **Range**, then enter 1 (in the first box) and 5 (in the second box). This will specify a count of valid responses for each item. Recall, our goal here is to count the number of valid responses given for each of the five items that compose the *Happiness* scale. A valid response for each item would be from 1 to 5. Therefore, to count the number of responses between 1-5, we enter these values in the **Range** cells.

8) Click **Add**.

9) The dialogue box will look similar to **Figure 4.67**.

10) Click **Continue**.

11) Click **OK** or **Paste**.

Figure 4.66   Count Occurrences of Values within Cases function

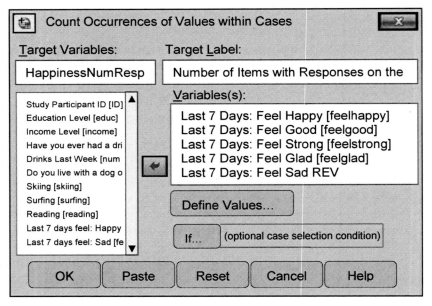

Figure 4.67   Specifying the **values to count** in the count values within cases

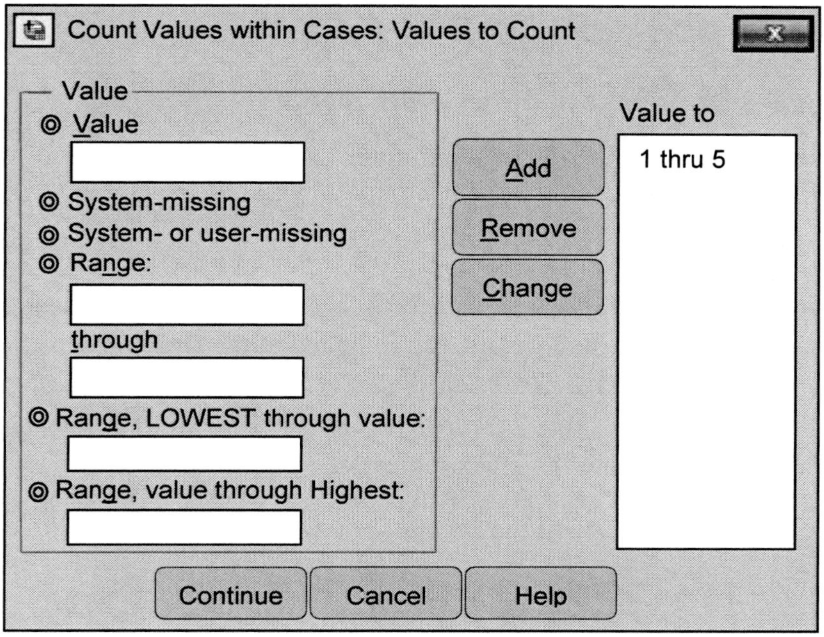

Now if you look at the end of the variables in the **Data View** portion of the SPSS database, as presented in **Figure 4.68** the new variable named *HappinessNumResp* will have been created. Most cases have a score of '5' (indicating all 5 items on the *Happiness* scale have a valid response). However, the case on the second line (**ID 2**) has a score of 4. This indicates that this case has valid responses for only four of the five items. If we look at all of the cells for **ID 2**, we can see that four cells have valid responses, while one is empty, regarding the five items that compose the Happiness scale. The final line (**ID 6**) has a score of '2' for *HappinessNumResp*, which indicates only two of the five items on the *Happiness* scale have valid responses. This is also indicated by the three empty cells (missing responses) for this line.

Figure 4.68   *HappinessNumResp* in the **Data View**

ID	feelhappy	feelsadREV	feelgood	feelstong	feelglad	HappinessNumResp
1.00	5.00	4.00	4.00	4.00	5.00	5.00
2.00	1.00	2.00	2.00	3.00		4.00
3.00	4.00	4.00	4.00	4.00	3.00	5.00
4.00	4.00	4.00	3.00	4.00	5.00	5.00
5.00	3.00	4.00	3.00	4.00	3.00	5.00
6.00	5.00	5.00				2.00

Next, we can conduct a **frequency analysis** of the variable *HappinessNumResp* to examine the number of items completed per study participant. To do so in SPSS one would go to:

1) **Analyze→ Descriptive Statistics→ Frequencies.**

2) In the dialogue box that opens (see **Figure 4.25**), move the variable *HappinessNumResp* from the left hand column into the right hand column.

3) Click **OK** or **Paste.**

SPSS Output 4.17 presents the statistical output for the **frequencies procedure**. The top smaller box indicates the number of cases included in the analysis, which indicates all 105 cases are *Valid* (included in the analysis) and 0 are *Missing*. The second box indicates the frequency of *the number of items with responses* for the five items composing the *Happiness* scale. This box indicates:

1) A frequency of 2 cases (1.9% of the sample) provided 2 valid responses

2) A frequency of 3 cases (2.9% of the sample) provided 3 valid responses

3) A frequency of 5 cases (4.8% of the sample) provided 4 valid responses

4) A frequency of 95 cases (90.5% of the sample) provided 5 valid responses

SPSS Output 4.17

Frequencies

Statistics

Number of Items with Valid
Responses on the Happiness
Scale

N	Valid	105
	Missing	0

		Frequency	Percent	Valid Percent	Cumulative Percent
Valid	2.00	2	1.9	1.9	1.9
	3.00	3	2.9	2.9	4.8
	4.00	5	4.8	4.8	9.5
	5.00	95	90.5	90.5	100.0
	Total	105	100.0	100.0	

We can compute the proportion of items with valid responses, thereby also reflecting the proportion of missing values, by dividing the number of completed items by the total number of scale items. To do this in SPSS go to:

1) **Analyze→ Transform→ Compute Variable**.

2) In the dialogue box that opens, enter a name for the new variable to indicate the proportion of missing data in the **Target Variable** cell. We have entered the name *HappinessPropMiss*, to reflect the variable that will indicate the proportion (*Prop*) of Valid/Missing (*Miss*) responses for the five items that form the *Happiness* composite scale.

3) Under the **Numeric Expression** cell, enter the coding, which is the name for the variable reflecting the number of completed items (*HappinessNumResp*), divided by the number of items, which is five (use the keypad to enter the coding).

4) The dialogue box should look similar to the **Figure 4.69**.

5) Click **OK** or **Paste**.

Figure 4.69   Computing the proportion of valid/missing data per case (top of box only)

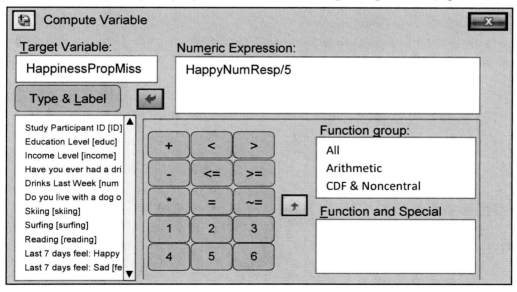

Figure 4.70 presents the new variable named *HappinessPropMiss* in the SPSS database reflecting the proportion of valid/missing responses. In this figure, on the left, when the variable *HappinessNumResp* indicates a '5' (reflecting 5 valid responses to the 5 items measuring *Happiness*), the cell to the right under the variable *HappinessPropMiss* indicates a score of '1.00,' which indicates 100% of data have been provided.

When the variable *HappinessNumResp* indicates a number less than '5,'which indicates missing data, the value for that case within the variable *HappinessPropMiss* is less than 1.00 (i.e., less than 100%). For example, where the variable *HappinessNumResp* indicates a '2' (reflecting 2 valid responses), the cell to the right under the variable *HappinessPropMiss* indicates a score of '.40,' which indicates 40% (2 of 5) of items have valid responses.

Figure 4.70 Noting the proportion of valid/missing data per case

	HappinessNumRes	HappinessPropMiss	
	5.00	1.00	
Valid responses:	4.00	.80	
5 of 5 (100%) →	5.00	1.00	← 100%
	5.00	1.00	
Valid responses:	5.00	1.00	
2 of 5 (40%) →	2.00	.40	← 40%

We can repeat the **frequencies procedure** for the variable *HappinessPropMiss* to reveal the proportion of valid/missing responses given among study participants for the five items forming the composite scale reflecting *Happiness*. **SPSS Output 4.18** presents the frequency counts for the variable *HappinessPropMiss*. We see the second box within this output indicates:

1) 2 cases (1.9%) provided valid responses for 40% (2 of 5) of items.

2) 3 cases (2.9%) provided valid responses for 60% (3 of 5) of items.

3) 5 cases (4.8%) provided valid responses for 80% (4 of 5) of items.

4) 95 cases (90.5%) provided valid responses for 100% (5 of 5) of items.

SPSS Output 4.18

Frequencies

Statistics

Proportion of Valid Responses
on the 5-Item Happiness Scale

N	Valid	105
	Missing	0

SPSS Output 4.18 (continued)

	Frequency	Percent	Valid Percent	Cumulative Percent
Valid .40	2	1.9	1.9	1.9
.60	3	2.9	2.9	4.8
.80	5	4.8	4.8	9.5
1.00	95	90.5	90.5	100.0
Total	105	100.0	100.0	

Summary. At this point we can now answer the question regarding the *proportion of missing data values* per study participant. Specifically, as reflected in **SPSS Output 4.18**, we have identified while over ninety percent (90.5%) of study participants provided complete data, 4.8% provided 80%, 2.9% provided 60%, and 1.9% provided 40% of valid responses to the five scale items.

Part B: The Proportion of Study Participants that have Missing Data Values

Next, we will examine the other dimension of missing data, which is the proportion of study participants that have missing data (Yes or No). This can be identified through a simple recode of the *HappyPropMiss* variable created in the last section. To perform this function in SPSS go to:

1) **Analyze→ Transform→ Recode into Different Variables**.

2) In the dialogue box that opens, under **Name**, enter a name for the new variable that will reflect if a study participant is missing a data point. We have entered the name *HappinessMissData*, to reflect the variable that will indicate if the study participant is missing data (*MissData*) on the *Happiness* scale.

3) Within the **Label** cell, enter a description of the new variable, such as *Is the Study Participant Missing Data?*

4) Click **Change**.

5) The dialogue box should look similar to **Figure 4.71**.

6) Click **Old and New Values**.

7) In the dialogue box that opens, under the **Old Value** column, click **Range** and enter the range of values that indicates missing data, which would be any value below 100% (within the *HappinessPropMiss* variable), so enter *0.0* to *0.999*, indicating any value from 0.0% to 99.9%.

Figure 4.71 Recode: *HappinessPropMiss* into *HappinessMissData*

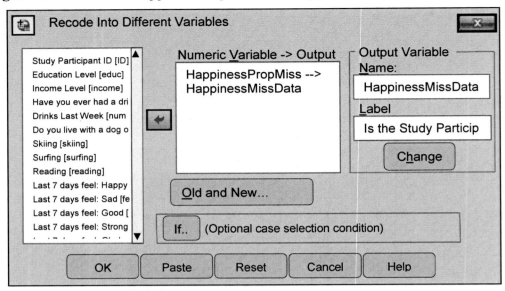

Figure 4.72 Old and New Values: Recoding into *HappinessMissData*

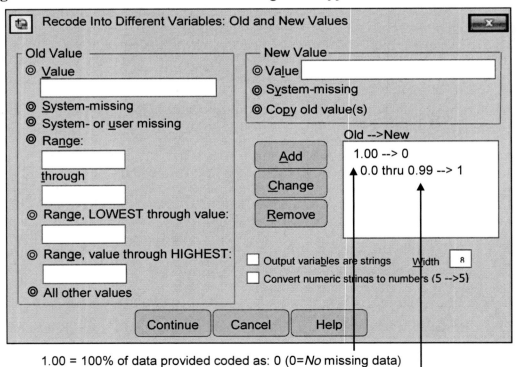

1.00 = 100% of data provided coded as: 0 (0=*No* missing data)

0.00-0.999 = 0-99.9% of data provided, coded as: 1 (1=*Yes* missing data)

8) Under the **New Value** column, click the toggle button for **Value** and enter the number '1,' which will indicate "Yes" (i.e., *Yes* there are missing data).

9) Click **Add**.

10) Under the **Old Value**, click **Value** and enter the value 1.0 reflecting no missing data, where 1.0 reflects 100% of data are present within *HappinessPropMiss*.

11) Under the **New Value** column, click the toggle button for **Value** and enter a '0' which will indicate "No" (i.e., *No* missing data) (presented in **Figure 4.72**).

12) Click **Continue**.

13) Click **OK** or **Paste**.

If you look at the end of the variables in the **Data View** portion of the SPSS database, as presented in **Figure 4.73**, the new variable named *HappinessMissData* will have been created. Notice in the figure, on the left, when the variable *HappinessPropMiss* indicates a value less that 100% (i.e., 0.0 tom 0.999), which reflects some data are missing, the cell to the right under the variable *HappinessMissData* indicates a value of '1.00,' which reflects a response of "Yes" (*Yes* there are some missing data).

Additionally, when there is a 1.0 for the variable *HappinessPropMiss* (indicating 100% of data are present), there is a value of '0.00' for the variable *HappinessMissData*, which indicates a response of "No" (*No* there is no missing data). Next, in the **Variable View** section of the SPSS database, enter the **Values** for the new variable *HappinessMissData*, where 1=Yes and 0=No.

Figure 4.73 *HappinessMissData* reflecting missing data (Yes or No) for each case

	HappinessPropMiss	HappinessMissData	
1.00=100% data →	1.00	.00 ←	.00=*No* Missing
.80=80% data →	.80	1.00 ←	1.00=*Yes* Missing

We can repeat the **frequencies procedure** for the variable *HappinessMissData* to reveal the proportion of study participants that have missing data. **SPSS Output 4.19** presents the frequency count for the groups coded as having and not having missing data points. The frequency counts correspond with the variable *HappinessPropMiss* where 95 (90.5%) study participants are coded as *No*, indicating there are no missing data (i.e., 100% of the five item scale have valid responses). The remaining 10 (9.5%) of study participants are coded as *Yes*, indicating some data are missing (i.e., less than 100% of the five item scale have valid responses).

SPSS Output 4.19

Frequencies

Statistics

Is the Study Participant Missing
Data on the Happiness Scale?

N	Valid	105
	Missing	0

Is the Study Participant Missing Data on the Happiness Scale?

		Frequency	Percent	Valid Percent	Cumulative Percent
Valid	No	95	90.5	90.5	90.5
	Yes	10	9.5	9.5	100.0
	Total	105	100.0	100.0	

Summary. At this point we can now answer the question regarding the *proportion of study participants* that have and do not have missing data values. Specifically, as reflected in **SPSS Output 4.19**, we have identified that 90.5% of study participants provided complete data (valid responses to each of the five items that form the *Happiness* composite scale), while 9.5% have some missing data.

In a Nutshell. Regarding the amount of missing data, the steps and procedures in this section have revealed:

1) Of 105 study participants, 10 (9.5%) have at least one missing data point (i.e., this is the proportion of study participants that have missing data points).

2) Of the 10 study participants with missing data values, five provided 80% of data, three provided 60% of data, and two provided 40% of data (i.e., this is the proportion of valid responses/missing data values per study participant).

4.4.3.3 Patterns of Missing Data

Missing data values may or may not evidence a pattern of *missingness*. Specifically, data might be missing in a pattern that suggests some factor is influencing which data points are missing. Commonly, researchers recognize three patterns of missingness, which will be presented in this section in Part A. A researcher must apply a method to examine which of these three patterns of missing data are present in a dataset. In Part B of this section, we will review a method of assessing the pattern of missing data in the dataset for study sample one. The pattern of missingness in a dataset can have dire implications. If the pattern of missing data indicates that missing values are related to some key factor, such as missing values being more likely among older study participants, specific steps may need to be taken to address this pattern of missingness. Therefore, we will now discuss what the patterns of missing data are, how to identify the pattern that exists in the dataset used in sample study one, and apply a method to account for the missing data values.

Part A: Types of Patterns of the Missing Data Values

In Part A we present an abbreviated description of the three commonly recognized patterns of missing data. A full discussion is beyond the scope of this book, although a great many sources do present these concepts in depth (Acock, 2005; Allison, 1999; Bennett, 2001; Roth, 1994). However, here we will mention the three patterns of missingness which are:

Missing completely at random (MCAR). The pattern MCAR indicates no patterns in the missing data. Essentially, this pattern is indicated when analysis reveals that the missing values are not significantly related to any other variables included in the study (Acock, 2005; Bennett, 2001; Roth, 1994). For example, analysis might indicate that study participants of older age are not more likely to evidence missing data values relative to study participants of a younger age.

Missing at random (MAR). First, you must keep in mind that the term missing at random is misleading. The pattern MAR may include data that are not completely missing at random. When missing data are MAR, the presence of missing data points may be significantly related to some other variable within the dataset. For example, an analysis of missing data patterns may reveal that males may be more likely to evidence missing data relative to females, in which case the missing data may be biased by gender. In short, the data are missing in a discernible pattern, but this pattern can be explained by another variable in the dataset, such as our example regarding gender. A data analyst may account for the pattern of MAR through incorporating the variable reflecting bias in analysis (e.g., gender) or simply reporting that the missing data are biased by that variable when reporting the study results.

Not missing at random (NMAR). NMAR (a.k.a., nonignorable nonresponse) indicates that the pattern of missingness *is* related to the variable that has the missing values (Allison, 2001). For example, NMAR might be indicated where a certain racial category is dramatically underreported, which might suggest that the study participants of that racial category tended not to report their racial identity. The NMAR pattern is often difficult to detect because the identification of this pattern is largely conceptual.

Furthermore, the pattern may be difficult to treat, as the source of the missingness may not be identifiable.

Part B: Assessing the Pattern of the Missing Data Values Within the Dataset

There are several effective methods that can be applied to assess the pattern of missing data in a dataset. For example, recently, programs have emerged that apply statistical software toward the analysis of patterns of missing data. The statistical software SPSS now offers a Missing Values Analysis (MVA) module that offers an effective means of detecting and addressing missing values. However, one of the most traditional methods for evaluating the randomness of missing values is to divide the sample into two groups that reflect those *with and without missing data values and compare them by important traits*.

This division into two groups may be based on one key variable or several key variables in a dataset. For example, in sample study one, we identified the groups with and without missing data based upon the key variable *Happiness*. In the last section (**4.4.3.2 Amount of Missing Data**) we identified these groups in **Part B: The Proportion of Study Participants that have Missing Data Values**. We then divided the study sample into the study participants that evidenced at least one missing valid response regarding the five items that form the *Happiness* composite scale ($n=10$), as well as the study participants that provided full data ($n=95$). This variable can be used to help identify the patterns of missing data in the dataset.

Specifically, we can now assess the patterns of missing data values through comparing if those with and without missing data (Yes or No) are related to essential characteristics of study participants, such as demographic traits. This is an important consideration as if the analysis indicates that the missing data are related to some important characteristic (age, gender, race, income, etc.,) the sample might be biased. A bias sample suggests the generalizability of the findings from the sample to the population may be compromised because certain profile of study participants is likely to be underrepresented. Therefore, in order to assess the patterns of missing data for sample study one, we will now examine if study participants with ($n=10$) and without ($n=95$) missing data values differ significantly by the demographic characteristics included in the dataset (i.e., *Education Level* and *Income Level*).

Education Level by Study Participants With/Without Missing Data Values. To first assess the pattern of missing data, we must compare the categorical variable *Education Level* by the categorical variable *Is the Study Participant Missing Data on the Happiness Scale?* Since we are comparing two variables, we need to use a bivariate test (*bi*=two, *var*=variable test). If we refer to the Bivariate Test Key in the beginning of section **4.6 Step 5: Bivariate Analysis**, we see a chi-square analysis should be used to examine the bivariate association between two categorical variables. Subsequently, we will use the chi-square procedure in this section to examine the relationship between the covariate variable *Education Level* and *Is the Study Participant Missing Data on the Happiness Scale?* The chi-square tests has certain assumptions regarding the data, which will be discussed in **Section 5.6.2**. However, to preliminarily conduct the chi-square analysis here in SPSS, go to:

1) **Analyze→ Descriptive Statistics→ Crosstabs**.

2) In the dialogue box that opens, use the arrow in the upper center to move the variable *Education Level* in the cell marked **Row(s)**.

3) Move the variable *Is the Study Participant Missing Data on the Happiness Scale?* in the cell marked **Column(s)** as presented in **Figure 4.74**.

Figure 4.74 Chi-Square Test: *Education Level* by *Missing Data*

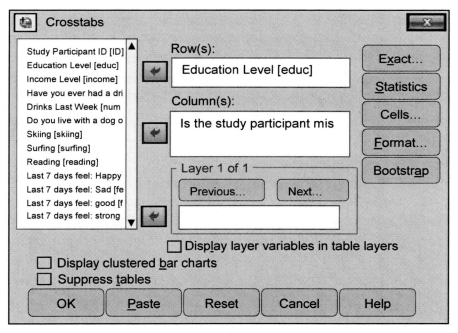

4) Either variable can go in the row or column, but many times the dependent variable/variable of interest is put in the column as a standard.

5) Click **Statistics**.

6) In that dialogue box that opens, check the **Chi-Square** box (see **Figure 4.75**).

7) Click **Continue**.

8) Click **Cells**.

9) In that dialogue box, check **Observed**, **Expected**, and **Row** (as in **Figure 4.76**).

10) Click **Continue**.

11) Click either **OK** or **Paste.**

Figure 4.75 Activating the **chi-square statistic** in the chi-square test

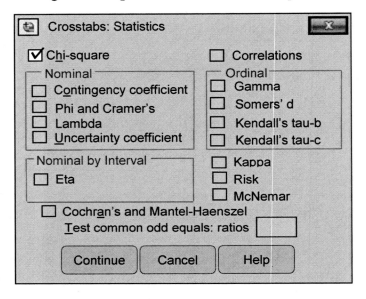

Figure 4.76 Activating **counts** and **percentages** in the chi-square test

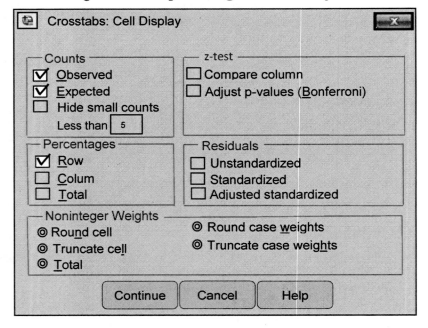

SPSS Output 4.20 presents the statistical output examining if there is a statistically significant relationship between *Education Level* and *Is the Study Participant Missing Data on the Happiness Scale?* To determine this, refer to the final box in the output labeled **Chi-Square Tests**. Within this box, the first line reads **Pearson Chi-Square**. If the value to the far right on this line, under the column **Asymp. Sig. (2-sided)**, is .05 or below ($p<.05$), then the test indicates a statistically significant relationship between variables. However, this column indicates the number **.816**, which is greater than .05 and indicates a statistically significant relationship does not exist. Therefore, there does not appear to be a bias by education level regarding study participants having and not having missing data points.

SPSS Output 4.20

Education Level * Is the Study Participant Missing Data on the Happiness Scale? Crosstabulation

			Is the Study Participant Missing Data on the Happiness Scale?		Total
			No	Yes	
Education Level	High School/GED	Count	29	3	32
		Expected Count	29.0	3.0	32.0
		% within Education Level	90.6%	9.4%	100.0%
	College	Count	27	2	29
		Expected Count	26.2	2.8	29.0
		% within Education Level	93.1%	6.9%	100.0%
	Graduate School	Count	39	5	44
		Expected Count	39.8	4.2	44.0
		% within Education Level	88.6%	11.4%	100.0%
Total		Count	95	10	105
		Expected Count	95.0	10.0	105.0
		% within Education Level	90.5%	9.5%	100.0%

Chi-Square Tests

	Value	df	Asymp. Sig. (2-sided)
Pearson Chi-Square	.406[a]	2	.816
Likelihood Ratio	.419	2	.811
Lin.-by-Lin. Assoc.	.113	1	.737
N of Valid Cases	105		

a. 3 cells (50.0%) have expected count less than 5. The minimum expected count is 2.76.

Sig. is above .05 ($p>.05$) at .816, reflecting the test is a not statistically significant

Income Level by Study Participants With/Without Missing Data Values. Next, to assess the pattern of missing data, we must compare the continuous variable *Income Level* by the dichotomous (two response categories) categorical variable *Is the Study Participant Missing Data on the Happiness Scale?* If we refer to the Bivariate Test Key in the beginning of **Step 5: Bivariate Analysis**, we see an independent-samples t-test analysis should be used to examine the bivariate association between a continuous variable and a dichotomous categorical variable. Subsequently, we will use the t-test procedure in this section to examine the relationship between the covariate variable *Income Level* and *Is the Study Participant Missing Data on the Happiness Scale* (Yes or No). The t-test incorporate assumptions we must test, which are addressed in the **Section 4.6.3** within sample study one. To conduct the independent-samples t-test in SPSS go to:

1) **Analyze→ Compare Means→ Independent Samples T-Test**.

2) Next, move the continuous variable (*Income Level*) from the left hand column into cell on the right under the label **Test Variable(s)**.

3) Move the categorical variable (*Is the Study Participant Missing Data on the Happiness Scale?*) into the cell labeled **Grouping Variable**.

4) Click **Define Groups** the dialogue box in **Figure 4.77** will open.

5) In **Group 1:** enter the number used to code group one for the categorical variable included in the test. Number '1' is entered in **Group 1** to reflect the first categorical group of study participants with missing data values.

6) In **Group 2:** enter the number used to code the other group for the categorical variable included in the test. Number '0' is entered in **Group 2** to reflect the second categorical group of study participants without missing data values.

7) Click **Continue** (The dialogue box should look similar to **Figure 4.78**).

8) Click **OK** or **Paste**.

Figure 4.77   Define groups in the independent-samples t-test

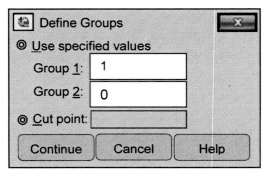

Figure 4.78   Dialogue box for the **Independent-Samples T-Test**

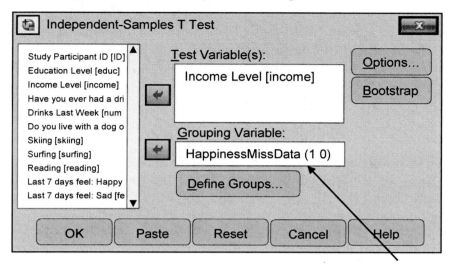

Is the Study Participant Missing Data on the Happiness Scale (Yes or No)

Output 4.21 presents the statistical output for the independent-samples t-test examining if there is a statistically significant difference by income level among study participants with and without missing data values.

SPSS Output 4.21

T-Test

Group Statistics

	Is the Study Participant Missing Data on the Happiness Scale?	N	Mean	Std. Deviation	Std. Error Mean
Income Level	Yes	10	214300.00	112033.775	35428.190
	No	95	201915.79	103890.763	10658.967

Independent Samples Test

		Levene's Test for Equality of Variances		t-test for Equality of Means					
		F	Sig	t	df	Sig. (2-tail)	Mean Diff.	95% CI of the Difference	
								Low	Up
Income Level	Equal Variances assumed	1.5	.23	.36	103	.72	1238	-57	81
	Equal Variances not assumed			.34	10.7	.74	1238	-69	94

Sig. is above .05 (*p*>.05) at .72, reflecting the t-test is <u>not</u> statistically significant

Within this output, note the box labeled **Independent-Samples t-test**. Within this box, refer to the line beginning with the term *Income Level* in the first column. If we follow this line over to the column labeled **Sig. (2-sided)**, we see the value for probability (statistical significance). A value greater than .05 ($p>.05$) indicates the variable relationship is not statistically significant. For example, because the value in this column is **.72**, this test indicates a statistically significant relationship does not exist between the variables reflecting *Income Level* and *Is the Study Participant Missing Data on the Happiness Scale?*

Summary. The bivariate tests reflect that data appear to be missing at random, as the study participants that did and did not evidence missing data values did not differ significantly by study variables reflecting essential characteristics, such as demographics (i.e., *Education Level, Income Level*).

4.4.3.4 Treatment of Missing Data

In this section we will first review some general information regarding the methods used to treat missing data points within a dataset. This is a rather brief discussion that is intended to make mention of several key points and procedures that are essential to know. A full discussion of the methods used to treat missing data would be extensive and is beyond the scope of this book. Therefore, in this discussion we will present the essentials of treating missing data, as well as apply a straightforward method of treating the missing data values within the dataset used for the sample studies presented.

There are a number of strategies for handling missing data that range from overly simplistic to very complex. Many of these methods can be conducted using standard statistical software packages (e.g., SAS, SPSS) and available freeware (e.g., Amelia, Norm), as well as packages specifically designed for particular types of analyses (e.g., Mplus for structural equation modeling). The proper method is often contingent upon the needs of the specific study.

The discussion of these methods to address the presence of missing data will focus upon the treatment of missing data among continuous variables. The treatment of missing values within categorical variables present more complex issues beyond the scope of this discussion, although a solid literature regarding methods of treating categorical missing data exists (e.g., Allison, 2001; Chen & Astebro, 2003; Graham, 2009). Several methods are mentioned in this section with references, so these techniques can be examined further. The following are a series of methods of accounting for missing data with a commentary on the procedure(s):

Deletion Methods. Deletion methods involve deleting cases based upon missing data points and are not recommended as a means of addressing significant challenges presented by missing data values. However, such methods warrant mention, as these techniques are the most common method used to deal with missing values. This may be due to the fact that they are the default method for treating missing data in such software as SPSS (specifically *Listwise deletion*). To put it briefly, deletion methods essentially ignore missing data rather than treating them. Unless there are particularly small degrees of missing data values, this is generally not a valid means of addressing challenges related to missing data. The two most frequently used deletion methods are:

Listwise deletion. In listwise deletion, cases with any missing data are simply deleted from the statistical analysis. This method of treating missing data is often referred to as *complete case analysis* (Pigott, 2001), as only cases with complete data are included in the analysis. As mentioned above this is the default procedure for many statistical programs, including SPSS. However, this method presents several issues, including the possibility that cases with complete data that are used in analysis represent a biased subsample of the total sample, which will in turn likely result in biased findings (Bennett, 2001).

Pairwise deletion. In pairwise deletion, also called *available case analysis* (Pigott, 2001), cases are excluded from analysis selectively on a variable by variable basis (Bennett, 2001; Roth, 1994). This method is problematic for a number of reasons, including the possibility that certain study participants (such as outlier scores) might be included in one analysis and excluded from another, which could easily bias the findings among several statistical tests.

Imputation Methods. Imputation strategies (which are preferable to deletion strategies) involve substituting a plausible value for the missing data point. Imputation methods travel along a wide spectrum from very basic (e.g., mean substitution) to quite sophisticated (e.g., multiple imputation). The attractive feature of imputation is that it facilitates the inclusion of all cases in analysis. The risk involved with imputation is that imputed values may be poor estimates of missing data points. Examples of imputation methods include:

Mean substitution. Mean substitution (at times Median substitution is also used) involves using the mean value score for valid cases to replace missing data points for a continuous variable. For example, let's suppose that out of 100 cases in a study, 90 cases provided data regarding their *annual household income* with a mean value of $100,000. Additionally, the remaining 10 cases failed to provide any data (i.e., missing data). Mean substitution would involve entering the mean score of the 90 valid cases (of $100,000) as the response for *annual household income* variable for the 10 cases that initially failed to provide any data.

Traditionally, mean substitution has been one of the most widely used but riskiest method of imputing a missing value. Indeed, the use of this method is strongly advised against by many experts (Allison, 2001; Bennett, 2001; Graham et al., 2003; Pallant, 2007). Among the many limitations of mean substitution is that this method assumes that the missing data are MCAR, where the missing data values are not likely to be significantly different from the non-missing data values, which is a dangerous assumption.

For example, in the case with missing data regarding *annual household income*, study participants often do not provide data when their income is either extremely lower or higher than average. Therefore, imputing the mean value (i.e., the average income) of the cases that did provide data for these missing cases is perhaps the worst method of estimating a missing value. First, this method does not provide a good estimate of what the missing value likely is. Second, the estimated value reduces the variance of scores in the sample. You can see this easily in our example regarding the variable annual household income. The missing scores are likely to be very low or high values. However, instead of letting those scores vary to these low and high levels, mean substitution would set the missing values as the average score.

Other approachable imputation methods worth examining include regression substitution (Graham et al., 2003) and hot-deck imputation (Bennett, 2001; MacCallum, Zhang, Preacher, & Rucker, 2002; Roth, 1994). However, where more advanced and sophisticated methods may be required, you may want to consider techniques such as expectation maximization (Bennett, 2001: Graham et al., 2003, p. 94), multiple imputation (Acock, 2005), and full information maximum likelihood (Enders & Bandalos, 2001; Graham et al., 1996; Olinsky, Chen, & Harlow, 2003).

Taking the Mean Score of Valid Responses (Treating the Missing Data Values in the Dataset Used In our Sample Study). It is important to realize that decisions made in data analysis relate to the ideology of the data analyst or statistician. Toward that, I am of the ideology where I would rather use the actual data provided by the study participant rather than enter estimated values, if a sufficient amount of data are provided. For example, regarding a 5-item scale like our *Happiness* measure, one might prefer to use the mean score of valid items where a sufficient proportion of items were answered, rather than estimate missing data values. This might be preferred because the items that have valid responses are actual study participant data. However, estimated values are not based upon actual data provided by the study participant.

For example, **we might require that 80% of items must have valid responses to be included in our analysis**. This translates into saying that at least four of the five items (80%) on our *Happiness* scale must have valid responses for a study participant to be included in analysis. That would mean that we would exclude study participants with less than 80% valid responses from analysis. It is commonly recognized that it is permissible to drop a small number of cases with missing data from analysis given a reasonable sample size. While there is no solid consensus among experts regarding the percentage of cases that might be removed before becoming problematic, cutoff points of 5% (Schafer, 1999) and 10% (Bennett, 2001) among somewhat large samples have been suggested. Since our sample is a relatively small 105 study participants, we might cautiously consider excluding a small number of study participants from the analysis due to missing data, certainly less than 5%.

Recall in our analysis, there were five cases (4.8%) that provided less than 80% of valid responses on the 5-item *Happiness* scale. These were the two cases (1.9%) that provided 40% and three cases (2.9%) that provided 60% valid responses on the *Happiness* scale (1.9% + 2.9% = 4.8%). If we apply our criteria of excluding cases with less than 80% of valid responses on the *Happiness* scale, we would exclude these five cases from analysis. Therefore, since this number represents less than 5% of the total sample, **we will remove these five cases from the analysis as a means of addressing missing data points**. However, we will save a copy of the old database with the five cases remaining. This will leave us with a sample of 100 study participants in our analysis rather than 105 study participants.

Although, keep in mind there were ten cases with missing data. Five were just excluded from analysis, but there are still five other cases that are to be included in analysis that still have a missing data value. These are the five cases that provided 80% of data (those who provided valid responses for 4 of 5 of the *Happiness* scale items). So now we need a plan to compute a *Happiness* score for these cases, even though one scale item is missing for each case.

One method to consider is computing a score for these study participants based on the 80% (four of the five items) of valid responses provided. Specifically, **we will compute a mean score based on the four items with valid responses, while ignoring the one item missing a valid response**. It is important to realize that the *Happiness* scale is computed through taking the mean score value of the five items on the scale. Many times scales are not computed in this manner. Thus, the method of accounting for a missing item through taking the mean value score of valid responses would only be useful where a scale is computed through taking the mean score of valid responses.

To compute the mean score of valid responses for the items that compose the *Happiness* scale in SPSS go to:

1) **Transform→ Compute Variable**.

2) In the dialogue box that opens, in the middle right hand section, under **Function Group**, highlight the option for **All** or **Statistical**.

3) Beneath that box, within the box marked **Functions and Special Variables**, select the term **Mean**.

4) Click the arrow to the center left of both boxes, which will put the syntax that reads **Mean (?,?)** in the top box under the term **Numeric Expression**.

5) From the list of variables on the left hand side, highlight and move each of the five items that compose the *Happiness* scale in between the parentheses next to the term **Mean**. Be sure each variable is separated by a comma.

6) In the **Target Variable** cell, enter the name for the composite score variable we are creating. You will see below that we have entered the name *Happiness*.

7) The dialogue box will look similar to **Figure 4.79**.

8) Click **OK** or **Paste**.

Figure 4.79 Computing the score for the 5-item composite *Happiness* scale

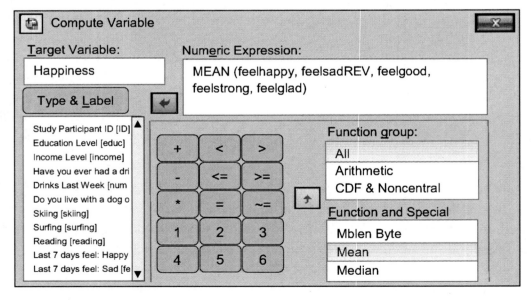

In the **Data View** section of the SPSS database, at the end of the variable list, the new variable named *Happiness* will appear as presented in **Figure 4.80**. Within this figure, we have lined up the five individual items that compose the *Happiness* scale, immediately before the new variable *Happiness*. You will see that on the second line for study participant **ID 2**, there are only four cells with valid responses, along with a missing value in the cell for the item *feelglad*. However, there is still a mean score for this case in the composite score for *Happiness* in the last column, which is the mean score for the four items with valid responses.

Of course this method is a trade-off. For example, through using the four valid responses for **ID 2**, we are using actual study participant data. However, there is a dimension of the 5-item scale missing. Specifically, the *feel glad* dimension of the scale is missing for the **ID 2**. We could use sophisticated imputation methods to estimate a value for the *feel glad* item for the case **ID 2**, which would facilitate all the dimensions of the scale being reflected. However, that imputed item would not be actual study participant data, but an imputed value.

Figure 4.80   The 5-item composite scale score reflecting *Happiness*

ID	feelhappy	feelsadREV	feelgood	feelstong	feelglad	Happiness
1.00	5.00	4.00	4.00	4.00	5.00	4.40
2.00	1.00	2.00	2.00	3.00		2.00
3.00	4.00	4.00	4.00	4.00	3.00	3.80
4.00	4.00	4.00	3.00	4.00	5.00	4.00
5.00	3.00	4.00	3.00	4.00	3.00	3.40

Summary. We treated the missing data in our sample study in two ways. First, we excluded cases where study participants provided less than 80% valid responses. This was only practical because small proportion (less than 5%) of cases in the sample fell into this category. Second, where study participants provided at least 80% valid responses, but had missing data, we used the mean score of the valid responses to compute the dependent variable score *Happiness*.

Even from our rather abbreviated discussion on the topic of missing data and the methods to treat missing data, it should be evident that these topics are far reaching issues in the world of data analysis. Therefore, we must state unequivocally that if possible and appropriate: **The best way to handle missing data is not to have any!**

4.4.4 Measurement Tools

When you first read that assessing study measurement tools is a check of data integrity, this statement might not seem to make sense. For instance, it may sound as if we are assessing the actual measurement tools, which might be completely independent of the study data. However, this would be an erroneous conclusion! The study measurement tools play a significant role in determining if data are appropriate for analysis. Specifically, it is necessary to assess the **reliability** and **validity** associated with each tool, toward assessing if measurement tools and the data are appropriate for statistical analysis.

In this section we will first present issues related to the **reliability (4.4.4.1)** of the study measurement tools. We will focus on perhaps the most commonly expressed dimension of reliability in quantitative research, which is internal consistency measured via the Cronbach's alpha. You will notice in this section that when assessing internal consistency reliability, we are not actually assessing the reliability of the measurement tool, but the reliability of the study data in the context of the measurement tool.

An examination of study measurement tool **validity (4.4.4.2)** will describe if a tool collected sufficient data to infer that the trait or construct being surveyed was the actual construct measured. Furthermore, validity may also describe if a trait or construct was measured sufficiently. Of course, if the study measurement tools do not seem to reflect the study constructs adequately, the study data will most certainly lack integrity. Therefore, when assessing if data are appropriate for analysis, an efficient assessment of the reliability and validity of the study measurement tools is an essential check of data integrity.

4.4.4.1   Scale Reliability

The **Reliability** of a measure can be described as the extent to which the instrument yields the same results on repeated measures. Within this process of yielding results, are the issues of *consistency* and *accuracy*. For example, an old bathroom scale that initially indicated you weighed 100 pounds, but then reported a weight of 200 pounds when you weighed yourself a second time a few minutes later, would be regarded as unreliable. Thus, a new bathroom scale that yielded the same score *consistently* would appear more reliable. In general, the less *variation* there is between measurements, the better the reliability.

In order to be truly reliable, this new bathroom scale must not only provide consistent results, but also *accurate* results. Specifically, if the scale reported a weight of 100 pounds consistently, but our accurate weight was 150 pounds, the scale would not be reliable. However, if you knew your weight was 100 pounds from a recent doctor visit, and the scale indicated this weight, the scale might appear to yield an accurate result.

Types of Reliability. There are several types of reliability, but not all types are needed for all studies. The type of reliability needed to be addressed within a particular study is often dependent upon the design and characteristics of the study. For example, when conducting a qualitative analysis, **interrater reliability** is often assessed. Interrater reliability is a means of demonstrating the equivalence or agreement among raters who are collecting data. For example, two raters observing an interview might separately code each time one senses a sign of anxiety in the interviewee. Interrater reliability might then be computed through determining the degree to which the codes indicating points of anxiety within the interviewee match between the two data gatherers. Often determining interrater reliability might involve computing a statistic, such as the Cohen's Kappa, which will reflect this degree of agreement. Since our sample study is purely quantitative, we would not need to compute interrater reliability in our analysis.

Quantitative studies often employ **test-retest reliability**. Test-retest reliability is intended to demonstrate the stability or consistency of a measure. To determine stability, a measure or test is administered initially, then again to the same subjects at a future date. Results from the test scores gathered at timepoint 1 and timepoint 2 are then compared and correlated as a measure of stability. For example, suppose we created a measure of self-esteem and administered it to 100 people. If the scale truly measured self-esteem reliably (and barring any events that might change their level of self-esteem), if we administered the same measure a month later the scores for each study participant should be roughly the same. In essence, the earlier score (timepoint 1) would be correlated with the later score (timepoint 2). However, since sample studies one and two involve a cross-sectional measurement of the dependent variable at one timepoint, we would not employ this measurement of reliability related to multiple timepoints in our analysis.

Perhaps the most ubiquitous measure of reliability in quantitative research is **internal consistency** reliability, which is most often expressed using the Cronbach's alpha. Typically, if a data analysis mentions only one rating of either the reliability or validity of a measure, it will be the Cronbach's alpha. A measure is said to be internally consistent to the degree that all scale items measure the same trait or construct. For example, regarding our measure of the dependent variable *Happiness*, the scale would be internally consistent if all the five items used to compute the total *Happiness* score were measuring the same construct (at least at an acceptable level). How do you know if all the items are measuring one trait or construct, such as *Happiness*? The answer is *Patterns*! The items within a scale will interact with each other in a certain pattern if a certain trait or construct is being measured. For example, note below the five items from our study survey used to compute the dependent variable *Happiness*, in SPSS.

Pattern indicating the highest level of *Happiness*: If a study participant has the highest level of *Happiness* possible, he or she will likely complete this scale in the pattern presented in **Figure 4.81** Specifically, he or she will select a 5 (Strongly Agree) for the positively stated items (Happy, Good, Strong, Glad) but a 1 (Strongly Disagree) for the negatively stated item (sad).

Figure 4.81   The pattern of responses indicating the highest level of *Happiness*

In the last 7 days I been:					
	Strongly Disagree	**Disagree**	**Neither Agree or Disagree**	**Agree**	**Strongly Agree**
Happy	1	2	3	4	(5)
Sad	(1)	2	3	4	5
Good	1	2	3	4	(5)
Strong	1	2	3	4	(5)
Glad	1	2	3	4	(5)

Pattern indicating the lowest level of *Happiness*: If a study participant has the lowest level of *Happiness* possible, he or she will likely complete this scale in the pattern presented in **Figure 4.82**. Specifically, he or she will select a '5'' (Strongly Agree) to the negatively stated item (sad) and '1' (Strongly Disagree) to the positively stated items (Happy, Good, Strong, Glad).

Figure 4.82   The pattern of responses indicating the lowest level of *Happiness*

In the last 7 days I been:					
	Strongly Disagree	**Disagree**	**Neither Agree or Disagree**	**Agree**	**Strongly Agree**
Happy	(1)	2	3	4	5
Sad	1	2	3	4	(5)
Good	(1)	2	3	4	5
Strong	(1)	2	3	4	5
Glad	(1)	2	3	4	5

Neither Extremely Happy nor Unhappy: If a study participant is neither extremely happy nor unhappy, he or she will likely complete the scale in the pattern of selecting the center values. Specifically, he or she will select a 3 (Neither Agree or Disagree) for the positively stated items (Happy, Good, Strong, Glad), as well as the negatively stated item (sad). Eventually, when enough study participants complete the 5-item *Happiness* measure, these patterns will begin to emerge in the data. When we perform a reliability analysis in the SPSS software, you will see that the software recognizes that a construct is being measured based on the presence of these patterns of responses.

No Pattern: Now consider the possibility of no pattern emerging. For example, in **Figure 4.83** note the selections made in the 5-item survey. The responses to these items are conflicting and suggests a single construct is not being measured. The study respondent selected both the highest level of being happy, sad, strong, and glad but the lowest level for good. Thus, these items seem to indicate that the construct *Happiness* is not being measured as it does not seem plausible that a person will experience feeling extremely *happy* and *sad* at the same time. Subsequently, the statistical reliability analysis will produce a low estimate of internal consistency for this scale, as this pattern of responses does not suggest a pattern where a single trait or construct is being measured.

Figure 4.83   The pattern of responses indicating the no pattern of *Happiness*

In the last 7 days I been:					
	Strongly Disagree	**Disagree**	**Neither Agree or Disagree**	**Agree**	**Strongly Agree**
Happy	1	2	3	4	⑤
Sad	1	2	3	4	⑤
Good	①	2	3	4	5
Strong	1	2	3	4	⑤
Glad	1	2	3	4	⑤

The Dance of the Survey Items. It may be helpful to think of the patterned set of responses presented in **Figure 4.81** and **Figure 4.82** as a coordinated dance between the survey items similar to the way dancers move in a coordinated unison as presented in the **Figure 4.84**. Notice as one dancer moves in a certain direction, the others also move in the same direction, similar to how items 1-5 on the *Happiness* scale move in a coordinated manner when one construct, i.e., *Happiness* is being measured.

Figure 4.84   Dancers moving in unison

Now consider when the survey items do not move in a patterned manner, as presented in **Figure 4.83**. If this were a dance it would look like the individuals presented in **Figure 4.85**, where the dancers use uncoordinated movements. It is important to remember that high internal consistency is reflected by the coordinated dance.

Figure 4.85   Dancers not moving in unison

Testing for Internal Consistency: Reliability Analysis. The level of internal consistency for a study measure can be assessed using a reliability analysis in the SPSS statistical software. To conduct this procedure in SPSS, go to:

1) **Analyze→ Scale→ Reliability Analysis**.
2) Move the five items that compose the *Happiness* scale from the left column into the right column labeled **Items**. Remember to use the reversed version of the *Feel Sad* item (*feelsadREV*), not the original (see **Figure 4.86**).
3) Click **Statistics**.
4) In the dialogue box that opens, click **Scale if item deleted** (as in **Figure 4.87**).
5) Click **Continue**
6) Click **OK** or **Paste**.

Figure 4.86 The **Reliability Analysis** for the 5-item *Happiness* composite scale

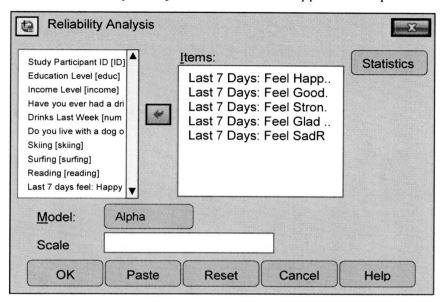

Figure 4.87 Activating the **scale if item deleted** function in reliability analysis

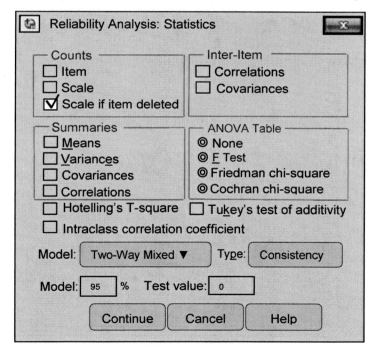

SPSS Output 4.22 presents the statistical output produced via the reliability analysis. The box labeled **Reliability Statistics** presents the value of the **Cronbach's alpha** rating which again, is a statistical measure of internal consistency for the items comprising the study measurement tool. The Cronbach's alpha can range from 0.00-1.00, with scores closer to 1.00 indicating higher internal consistency. There is a general agreement that the Cronbach's alpha rating must be at least 0.70 to be considered acceptable, although there are many reports that recommend a higher number with acceptable alpha values ranging from 0.70 to 0.95 (Bland, 1997; DeVellis, 2003; Nunnally, 1994). However, there is also a rule where if an alpha value is too high, it may suggest that some items are redundant, so a maximum alpha value of 0.90 has been recommended (Streiner, 2003). Within this statistical output, we see that the Cronbah's alpha rating for our *Happiness* scale is an acceptable .887.

The next box, labeled **Total Item Statistics** presents several values. The last column to the right labeled **Cronbach's Alpha if Item Deleted** indicates what the Cronbach's alpha would be for the scale if that item was deleted. You can see for all four of the individual items, the values indicate deleting that item would bring the Cronbach's alpha lower (below.877). Furthermore, deleting the fifth item *Feel Glad* would only increase the alpha marginally (to .878). Therefore, it seems that all five items are contributing to the measurement of a single item or trait and seem appropriate for the scale.

SPSS Output 4.22

Reliability

Reliability Statistics

Cronbach's Alpha	N of Items
.887	5

A Cronbach's alpha of .70 or above indicates an ← acceptable level of internal consistency

Item-Total Statistics

	Scale Mean if Item Deleted	Scale Variance if Item Deleted	Corrected Item-Total Correlation	Cronbach's Alpha if Item Deleted
Last 7 Days: Feel Happy	10.92	8.276	.723	.855
Last 7 Days: Feel Good	11.16	8.984	.747	.842
Last 7 Days: Feel Strong	11.04	9.817	.729	.848
Last 7 Days: Feel Glad	11.06	11.148	.602	.878
Last 7 Days: Feel Sad(REV)	10.50	8.818	.800	.828

The **Scale if Item Deleted** function is more useful when the Cronbach's alpha is unacceptably low. Specifically, when the Cronbach's alpha is unacceptable low, the **Scale if Item Deleted** function will inform you of which individual items to remove to increase the Cronbach's alpha rating of that scale. For example, please note **SPSS Output 4.23**, which presents a second reliability analysis, where we have removed the reversed version of the item *Feel Sad* (*FeelSadREV*) and replaced it with the original scale item *Feel Sad*. Now a higher score on the *Feel Sad* item conflicts with the other four items toward measuring *Happiness*, which will negatively impact internal consistency.

Within this output, the second box informs us that the Cronbach's alpha is an unacceptably low at .227. Within the third box labeled **Item-Total Statistics,** the final line reports that if the item *Feel Sad* (highlighted) is deleted, the Cronbach's alpha would increase to .828 (within the column **Cronbach's Alpha If Item Deleted**). This is an example of how a low Cronbach's alpha rating might be addressed to increase the level of scale internal consistency.

SPSS Output 4.23

Reliability

Reliability Statistics

Cronbach's Alpha	N of Items
.227	5

Cronbach's Alpha Is Unacceptably ← Low at Less than .70

Item-Total Statistics

	Scale Mean if Item Deleted	Scale Variance if Item Deleted	Corrected Item-Total Correlation	Cronbach's Alpha if Item Deleted
Last 7 Days: Feel Happy	10.58	2.084	.560	-.568[a]
Last 7 Days: Feel Good	10.82	2.513	.567	-.425[a]
Last 7 Days: Feel Strong	10.70	3.020	.521	-.244[a]
Last 7 Days: Feel Glad	10.72	3.396	.553	-.163[a]
Last 7 Days: Feel Sad	10.50	8.818	-.800	.828

a. The value is negative

If this Item Is deleted the Cronbach's Alpha would be an acceptable ($\geq$.70) .828

4.4.4.2 Scale Validity

Validity is the other key area that must be assessed when examining the quality of a measurement tool. The **Validity** of a measure can be described as the extent to which a measurement tool measures the specific trait or construct it is supposed to measure. One could say, a rating of **reliability**, such as the Cronbach's alpha measure of **internal consistency**, indicates: *Is one trait or construct being measured reliably?* (Regardless of what that trait or construct is called). And a measure of **validity** might tell us: *What is that one trait or construct being measured reliably?* (What is it called?).

For example, in the last section the Cronbach's alpha rating of internal consistency for the five items that compose the *Happiness* measurement tool was .878, which indicated that one trait or construct is being measured reliably. However, what is that trait or construct? We are calling the global trait being measure by the five items *Happiness*. However, just because we named the scale *Happiness that* does not guarantee that this is the trait actually being measured. How can we be sure we are not in fact measuring a similar trait, such as optimism? To assess the validity of the scale we need to apply certain criteria to the measure.

No Reliability, Means No Validity for You! As you may have deduced from the above information, reliability and validity are not independent qualities of a measurement tool, but are interrelated. For example, a measurement tool that is not reliable cannot be valid. Recall, if a measurement tool is not reliable, the scores resulting from that measure can be both inconsistent and inaccurate. Thus, after determining that a tool provides inconsistent and/or inaccurate scores (the tool is not reliable), it just makes sense that we would not consider calling such a tool valid. For example, if the Cronbach's alpha rating of the five items that compose the *Happiness* scale was unacceptably low (e.g., below .70), it would not have been appropriate to move onto this examination of validity to identify the name of the trait or construct being measured. This is largely because the reliability analysis (Cronbach's alpha rating) indicated a low level of internal consistency (>.70), which conveyed the message that the items do not measure anything (reliably).

No Validity = Maybe Reliability for You? Perhaps somewhat oddly, a measurement tool can be reliable without being valid. Specifically, a measurement tool can be consistent and accurate (i.e., reliable), but not have the valid criterion to be an indicator of the specific trait or construct targeted. For example, suppose a researcher was under the impression that when people are happy, they eat a 16 ounce cup of ice cream. Furthermore, the happier they are over the course of a week, the more 16 ounce cups of ice cream they eat.

Subsequently, that researcher changes the measurement tool of *Happiness* from our 5-item scale to indicate level of *Happiness*, to counting the number of 16 ounce cups of ice cream one eats over the course of a week to reflect their level of *Happiness*. The more 16 ounce cups of ice cream eaten, the *Happier* the person is said to be. In terms of reliability, using a 16 ounce cup of ice cream as a measurement tool would be highly reliable as the measure is both consistent and accurate. The measure is consistent and accurate because at each measurement point, the same amount (16 ounces) is eaten.

However, the measurement tool (16 oz. cup of ice cream) might not be valid, or more specifically, not a valid measure of *Happiness*. You could make the argument that people eat ice cream for many reasons, when they are happy, sad, or for the most part neither. Therefore, ice cream does not reflect the presence of the trait *Happiness*. Thus, although highly reliable, using the 16 oz. cup of ice cream as the measurement tool for the trait of *Happiness* might not be valid.

Types of Validity. There are many aspects of validity that are applied to a measurement tool, few of which employ statistical tests. Examining the validity of a tool is largely a conceptual task. This means assessing validity requires a lot of thought. In order to indicate a measurement tool is valid, you must provide a sufficient amount of evidence from which validity can be inferred. When using a standardized measurement tool in research, there are often peer-reviewed articles and research papers that report on the aspects of validity that you can cite.

Typically, the first type of validity assessed is **face validity**. Face validity addresses if an instrument *seems* to measure the trait or construct it is intended to measure. For example, when we assess the five items that compose the *Happiness* scale, we see that the items ask study participants about feeling happy, good, glad, strong, and sad. Each of these traits *appears* to indicate the presence (or lack of presence regarding sad) of the trait *Happiness*, so we might consider the measure has face validity.

Content validity refers to the degree to which a measurement tool has a pool of items that adequately reflect the dimensions of the trait or construct being measured. Content validity is based on the judgment of the rater. Thus, many times to evaluate the content validity of a measurement tool, an entire panel of experts on the trait or construct being measured is convened to assess the instrument. For example, if a researcher created a new measure of clinical depression, he or she might convene a panel of experts on the measurement of depression to be sure the items address the appropriate dimensions of depression and not a related construct, such as hopelessness.

We are not employing an expert panel to assess our study tool measuring *Happiness*, which does not bolster tool validity. Furthermore, you could say the measurement tool is lacking in an appropriate number of items, as it is dubious that a multidimensional construct like *Happiness* could be measured using only five items. Thus, in terms of validity, the measurement tool used to measure the dependent variable in sample studies one and two might benefit from some enhancement.

Summary. In the section regarding measurement tools, we discussed the broad topics of reliability and validity. Although we did not go into great detail, it is important to realize that the measurement tools selected for a study are extremely important and should be a central focus of study design and analysis. Briefly, if the constructs are not measured accurately, then the entire study will be misguided. Therefore, before approaching any phase of descriptive or inferential statistical analysis, a rigorous assessment of the validity and the reliability of the measurement tools is essential.

4.5   Step 4: Univariate Analysis

Univariate analysis explores each variable in a dataset separately. This is reflected in the term univariate analysis as the prefix *uni* suggests the number one and the root *var* suggests variable, indicating the analysis of one variable. Univariate analysis is also referred to as **descriptive analysis**, as these statistics essentially *describe* variables. It may seem logical to conduct an initial univariate analysis where individual variables are described, prior to inferential analysis where variables are related to one another.

Univariate analysis commonly involves reporting measures *of central tendency* and *dispersion*. Measures of central tendency involve describing how the values within a distribution of scores *tend* to be clustered *centrally*. The most common measure of central tendency is the **mean** score, which is simply the arithmetic average of a set of scores (where the values of a number of items are summed, then that summed number is divided by the number of items summed). If a distribution of scores in non-normal, the **median** (the middlemost numerical value separating the higher half of a distribution of scores from the lower half) or **mode** (the value that appears most often in distribution of scores) might be a more representative measure of central tendency. If a distribution of scores is perfectly normal, the mean, median, and mode will intersect.

Another set of values presented in univariate analysis, involve measures of *dispersion*, which describe how scores tend to be dispersed in a distribution. Common measures of dispersion include the **range** (the highest − lowest value), which reflects the distance between the lowest and highest value, as well as the **minimum and maximum** scores (the lowest and highest values in a distribution of scores). A more sophisticated measure of dispersion is the **standard deviation**. The standard deviation portrays the dispersion of values around the mean. Specifically, the standard deviation reflects the average variability (or spread) of the scores from the mean value.

Please see the **Figure 4.88**, which presents the *Univariate Test Key*, which is an abbreviated key that describes how univariate analysis is typically conducted in a data analysis study. In the first column (labeled **Variable**) we see there are two types of variables that need to be presented via univariate analysis, **Categorical** (nominal and ordinal level variables) and **Continuous** (ratio and interval level variables). In the center column (labeled **Test to Use**) we see the **Frequencies** procedure in SPSS will be used for both types of variables. However, in the final column (**Values to Report**) we see that we will compute and report different statistical values for the categorical (number and percentage by category) and continuous (mean, standard deviation, and minimum/maximum score).

Figure 4.88   Univariate Test Key

Variable	Test to Use (Values to Report)
Categorical	Frequencies (Number & Percentage)
Continuous	Frequencies (Mean, Std. Deviation, Min/Max)

Plugging in the Study Variables. Please refer to **Figure 4.89**, which presents the *Univariate Test Key* with the study variables plugged in. Once we have identified the variables used in the sample study as categorical (*Education Level, Do You Live with a Dog or Cat?*) and continuous (*Income Level, Happiness*), it is quite easy to plug them into the *Univariate Test Key* toward identifying the statistical values to compute and report using the **frequencies procedure** in SPSS.

Figure 4.89   Univariate Test Key (Study Variables Plugged In)

Variable	Test to Use (Values to Report)
Categorical	Frequencies (Number & Percentage)
Education Level	Frequencies (Number & Percentage)
Lives w/a Dog/Cat	Frequencies (Number & Percentage)
Continuous	Frequencies (Mean, Std. Deviation, Min/Max)
Income Level	Frequencies (Mean, Std. Deviation, Min/Max)
Happiness	Frequencies (Mean, Std. Deviation, Min/Max)

4.5.1 Categorical Variables

Computing univariate statistics for categorical variables is rather simple. In fact, most statistics textbooks often only describe how to report univariate statistics for continuous variables. However, it is useful to keep in mind that when describing a categorical variable, one should present the number and percentage of study participants that fall into each response category. For example, one should present the number and percentage of study participants that fall into the response categories for the covariate variable *Education Level* (High School/GED, College, and Graduate School). These values are produced in SPSS by going to:

1) **Analyze→ Descriptive Statistics→ Frequencies**.

2) In the dialogue box that opens (displayed in **Figure 4.25**), move the categorical study variables (*Education Level, Do You Live with a Dog or Cat?*) from the left hand column into the right hand column.

3) Click **OK** or **Paste**.

SPSS Output 4.24 presents the statistical output for the frequencies procedure. The first box labeled **Statistics** reports the number of cases with valid and missing values, which indicates 100 valid cases and 0 missing for each variable, which means both variables have full data and no missing data points. The next box (labeled **Frequency Table**) presents the frequency counts reflecting how many study participants are in each category within each variable. For example, regarding the variable *Education Level*, among the 100 study participants, 32 (32%) are at the High School/GED level, 29 (29%) are at the College level, and 39 (39%) are at the Graduate school level. Regarding the variable *Do You Live with a Dog or Cat?* of the 100 study participants, 47 (47%) live with a dog and 53 (53%) live with a cat.

SPSS Output 4.24

Frequencies

Statistics

		Education Level	Live with a dog or a cat?
N	Valid	100	100.0
	Missing	0	0

SPSS Output 4.24 (continued)

Frequency Table

Education Level

		Frequency	Percent	Valid Percent	Cumulative Percent
Valid	High School/GED	32	32.0	32.0	32.0
	College	29	29.0	29.0	61.0
	Graduate School	39	39.0	39.0	100.0
	Total	100	100.0	100.0	

Live with a dog or a cat?

		Frequency	Percent	Valid Percent	Cumulative Percent
Valid	Dog	47	47.0	47.0	47.0
	Cat	53	53.0	53.0	100.0
	Total	100	100.0	100.0	

4.5.2 Continuous Variables

As stated earlier, computing univariate statistics for continuous variables largely involves measures of central tendency and dispersion. Recall in **Figure 4.88** the *Univariate Test Key* noted the most commonly cited measure of central tendency, which is the *mean* value, as well dispersion, which are the *standard deviation* and *minimum/maximum* values. To produce these univariate statistics for our continuous study variables, in SPSS go to:

1) **Analyze→ Descriptive Statistics→ Frequencies**.

2) In the dialogue box that opens (displayed in **Figure 4.25**) use the center arrow to move the continuous variables (*Income Level*, *Happiness*) from the left column into the column on the right.

3) Click **Statistics**.

4) In the dialogue box that opens, click **Mean, Median, Mode**, (measures of central tendency), as well as **Std. deviation, Minimum,** and **Maximum** (measures of dispersion) as presented in **Figure 4.90**.

5) Click **Continue**.

6) Click **OK** or **Paste**.

Figure 4.90 Activating measures of central tendency and dispersion

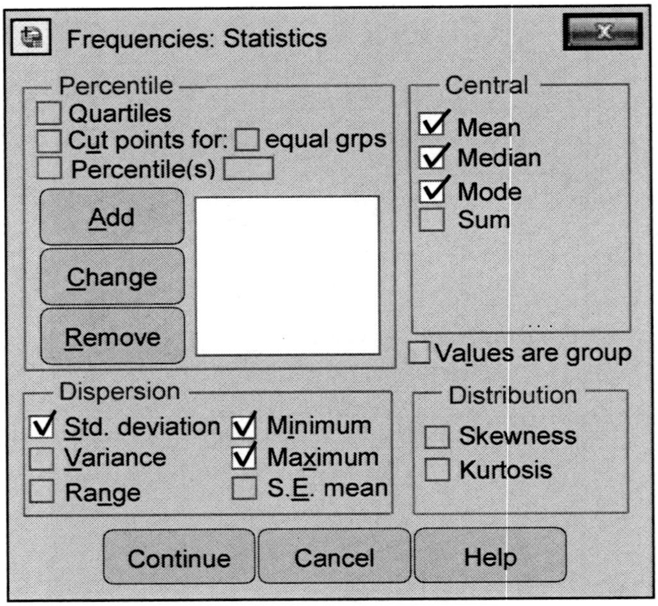

SPSS Output 4.25

Frequencies

Statistics

		Income Level	Happiness
N	Valid	100	100
	Missing	0	0
Mean		197250.00	2.7340
Median		178000	2.6000
Mode		10000	2.40
Std. Deviation		103336.791	.75535
Minimum		100000	1.20
Maximum		495000	4.40

SPSS Output 4.25 presents the statistical output for this frequency procedure. The first box (labeled **Statistics**) in the output reports the number of cases with valid and missing values. Again, this box indicates that for both variables there are complete data for all 100 study participants. That same box presents the values for *Income Level* reflecting the mean (21.72), median (21.00), Mode (21.00), standard deviation (2.55), and minimum (18.00) /maximum (26.00) values. Furthermore, in the same box we see the values for *Happiness* reflecting the mean (2.73), median (2.60), Mode (2.40), standard deviation (.76), and minimum (1.20)/maximum (4.40) values. The following two boxes present the frequency counts for each individual value within each variable.

4.6   Step 5: Bivariate Analysis

After employing Univariate Analysis to examine the variables by themselves, we may now want to look at how the variables relate to one another using Bivariate Analysis. Bivariate Analysis involves looking at how two variables at a time relate to one another. Indeed, if we break down the term bivariate we see the roots of the term indicate looking at two variables where the term *bi* indicates two and *var* indicates variable. Recall, we want to include all study variables related to the dependent variable at a statistically significant level ($p<.05$) in the multivariate regression model to statistically control for the influence of those variables in the final stage of analysis. Therefore, a primary goal of bivariate analysis is not just to test the association between two variables, but also identify which study variables are related to the dependent variable, so they may be included as predictors in the final multivariate regression model conducted in **Step 6: Multivariate Analysis**.

For example, in sample study one, a goal of bivariate analysis would be to identify which study predictor variables (*Education Level, Income Level, Do You Live with a Dog or Cat?*) are significantly related to the dependent variable *Happiness* on a one on one level (i.e., bivariate level). Each study predictor variable related to the dependent variable at a statistically significant level, will be included in the multiple linear regression model examining *Happiness*. Any predictor variables not significantly related to the dependent variable will be excluded from the multivariate analysis.

Figure 4.91 presents the *Bivariate Test Key*, which describes the appropriate statistical procedure to test the association between two variables, will depend on the structure of each variable. **Figure 4.92** presents the *Bivariate Test Key* with the study variables plugged in. The first line instructs if Variable 1 is categorical with a two category response, such as *Do You Live with a Dog or a Cat?* (Response: Dog, Cat) and Variable 2 is continuous, such as *Happiness* (Response: 1-5), then an **Independent Samples T-Test** would be the appropriate statistical procedure to test this association. The second line within the figure describes that if Variable 1 is categorical with three or more categories, such as *Education Level* (Response: High School/GED, College, or Graduate School) and Variable 2 is continuous, such as *Happiness* (Response: 1-5), then a **One-Way ANOVA** would be the appropriate statistical test to examine this association. The third line instructs if Variable 1 is continuous, such as *Income Level* (Response: $100,000-$500,000) and Variable 2 is continuous, such as *Happiness*

(Response: 1-5), then a **Correlation** would be the appropriate statistical test to examine this association. Sample study one does not require the chi square analysis listed in the fourth line, but this test will be reviewed in detail in sample study two.

Figure 4.91   Bivariate Test Key

Variable 1	Variable 2	Test To Use
Categorical (2 categories)	Continuous	Independent Samples T-Test
Categorical (≥3 categories)	Continuous	One-Way ANOVA
Continuous	Continuous	Correlation
Categorical	Categorical	Chi-Square

Figure 4.92   Bivariate Test Key (Study Variables Plugged In)

Variable 1	Variable 2	Test To Use
Categorical (2 categories)	Continuous	Independent Samples T-Test
Live w/Dog/Cat (Dog, Cat)	Happiness (1-5)	Independent Samples T-Test
Categorical (≥3 categories)	Continuous	One-Way ANOVA
Education (HS/G, C, GS)	Happiness (1-5)	One-Way ANOVA
Continuous	Continuous	Correlation
Income ($100k - $500k)	Happiness (1-5)	Correlation
Categorical	Categorical	Chi-Square

4.6.1   One-Way ANOVA

The Bivariate Test Key indicates a **One-Way ANOVA (Analysis of Variance)** is the appropriate statistical procedure examining the relationship between a categorical variable with three or more response categories with a continuous variable, such as the categorical covariate variable *Education Level* (Response Categories: High School/GED, College, Graduate School) with the continuous dependent variable *Happiness* (Response Range: 1-5). This procedure will indicate if there is a statistically significant difference in mean *Happiness* scores among study participants with a High School/GED, College, or Graduate School level of education.

The **One-Way ANOVA** procedure has certain assumptions, including **independence of observations** (there is no relationship between the study participants in each group or between the groups themselves, such as there are different participants in each group with no participant being in more than one group), **no significant outliers** (the distributions of scores in the continuous variable should not contain significant outliers scores), **normality** (the scores within the continuous variable are normally distributed), and **homoscedasticity** (homogeneity of variances). These assumptions were examined in **Step 3: Checks of Data Integrity**.

It is important to realize that the number of study participants within each response category within the categorical variable included in the ANOVA is also an important consideration as this will impact statistical power (Kraemer & Thiemann, 1987). In order to detect a medium-to-large effect size, a minimum of *30 participants per cell* has been suggested (Cohen, 1988). As the number of study participants within each response category drops, the power to detect statistically significant differences between groups also wanes. Regarding the **One-Way ANOVA**, 14 study participants per cell will provide approximately 80% power to detect a large effect size (.50), while seven study participants per cell will provide approximately 50% power to detect a large effect size (.50). If a small number of study participants per cell presents an issue, consider a **Kruskal-Wallis *non-parametric ANOVA***, which is also available in SPSS. To conduct the One-Way ANOVA in SPSS go to:

1) **Analyze→ Compare Means→ One-Way ANOVA**.

2) In the dialogue box that opens, enter the continuous variable (the dependent variable *Happiness*) from the left hand column into the right labeled **Dependent List**, using the top arrow button.

3) Enter the categorical variable, the covariate predictor variable *Education Level*, into the lower box labeled **Factor** (as presented in **Figure 4.93**).

Figure 4.93   One-Way ANOVA: *Happiness* by *Education Level*

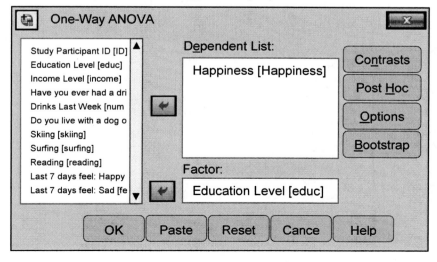

4) Click **Post Hoc**.

5) In the dialogue box that opens, click **Bonferroni** (see **Figure 4.94**).

6) Click **Continue**.

Figure 4.94   Activating the **Bonferroni post hoc test** in the One-Way ANOVA

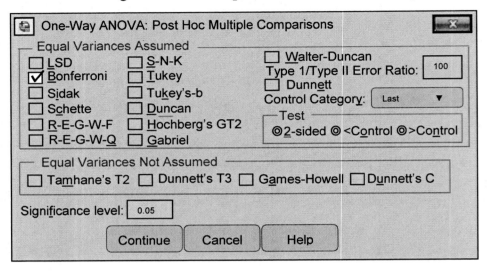

7) Click **Options**.

8) In the dialogue box that opens, click **Descriptive**, as presented in **Figure 4.95**.

Figure 4.95   Activate the **descriptive** in the One-Way ANOVA

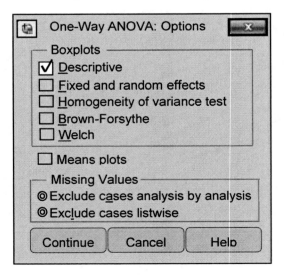

9) Click **Continue**.

10) Click **OK** or **Paste**.

Interpreting the Statistical Output: One-Way ANOVA. **SPSS Output 4.26** presents the statistical output for the **One-Way ANOVA** procedure. In this section we use this statistical output to identify the three dimensions of the relationship between the dependent variable *Happiness* and the covariate variable *Education Level*. Specifically, we will present the relationship dimensions of significance, directionality, and magnitude in three separate parts (Part A, Part B, and Part C, respectively).

Part A   Relationship Dimension 1: Significance

Note the second box labeled **ANOVA** within the output. In the final column labeled **Sig.**, the level of probability (statistical significance) for the overall test is indicated, which is **.000**. Thus, the overall model for the **One-Way ANOVA** procedure is statistically significant ($p<.05$). Therefore, we will continue to examine the results of the test.

Part B   Relationship Dimension 2: Directionality

Next, refer to the top box labeled **Descriptives**, which presents the mean scores and standard deviation values for the dependent variable *Happiness* by each response category within the variable *Education Level*. We see that the mean score for *Happiness* for study participants with a High School/GED (*M*=2.17, *SD*=.69) and College (*M*=2.70, *SD*=.61), seem to be much lower relative to that of Graduate School (*M*=3.23, *SD*=.55). Also, refer to the third box labeled **Multiple Comparisons** under the term **Post Hoc Tests**. These values will indicate which mean score is significantly different than another. The first column labeled **(I) Education Level**, presents each category separately. The second column labeled **(J) Education Level**, lists the other response categories. The third column labeled **Mean Difference**, presents the differences between the mean scores relative to the category listed in the first column. The asterisk next to the mean difference value indicates a statistically significant difference, in tandem with the forth column labeled **Sig.**, which presents probability levels.

In the first row of the **Multiple Comparisons** box, we see that mean value *Happiness* scores for the study participants with a High School/GED education level, are significantly lower relative to those with a College (-.52780) and Graduate School (-.05689) level of education. In the next row, we see that mean value *Happiness* scores for the study participants with a College education level, are significantly higher relative to those with a High School/GED (.52780) and lower than Graduate School (-.52909) level of education. Lastly, the final row describes mean value *Happiness* scores for study participants with a Graduate School education level, are significantly higher relative to a High School/GED (1.05689) and College (.52909) level of education.

SPSS Output 4.26

Oneway

(2ⁿᵈ) Directionality:

Higher education level = Higher *Happiness*

Descriptives

Happiness

	N	Mean	SD	SE	95% CI for the Mean		Min	Max
					Lower	Upper		
High School/GED	32	2.17	.70	.12	1.92	2.42	1.2	4.2
College	29	2.70	.61	.11	2.47	2.93	2.0	4.4
Graduate School	39	3.23	.55	.09	23.05	3.40	2.0	4.4
Total	100	2.73	.76	.08	2.58	2.88	1.2	4.4

ANOVA

(1ˢᵗ) Significance (Model):

Statistically significant

(p<.05)

Happiness

	Sum of Squares	df	Mean Square	F	Sig.
Between Groups	19.69	2	9.85	25.96	.000
Within Groups	36.79	97	.38		
Total	56.48	99			

Post Hoc Tests

(1ˢᵗ) Significance (Mean Scores):

Statistically significant (p<.05)

Multiple Comparisons

Dependent Variable: Happiness

Bonferroni

(Mean score differences)

(I) Education Level	(J) Education Level	Mean Difference (I-J)	Std. Error	Sig.	95% CI for the Mean	
					Lower	Upper
High School/GED	College	-.52780*	.16	.004	-.91	-.14
	Graduate School	-1.05689*	.15	.000	-1.41	-.70
College	High School/GED	.52780*	.16	.004	.14	.91
	Graduate School	-.52909*	.15	.002	-.90	-.16
Graduate School	High School/GED	1.05689*	.15	.000	.70	1.41
	College	.52909*	.15	.002	.16	.90

*The mean difference is significant at the 0.05 level.

Part C   Relationship Dimension 3: Magnitude

The One-Way ANOVA procedure in SPSS does not have an option to estimate the effect size, but the related ANCOVA procedure will produce an effect size estimate known as the partial Eta squared statistic, which can be produced in SPSS through the path:

1) **Analyze→ General Linear Models→ Univariate.**

2) Within the dialogue box that opens, enter the same continuous variable used within the **One-Way ANOVA**, the dependent variable *Happiness*, into the top cell under the term **Dependent Variable.**

3) Enter the categorical variable from the **One-Way ANOVA**, the covariate predictor variable *Education Level* into the cell under the term **Fixed Factor(s).**

4) The dialogue box will look similar to the **Figure 4.96.**

Figure 4.96   Producing an effect size using a general linear model

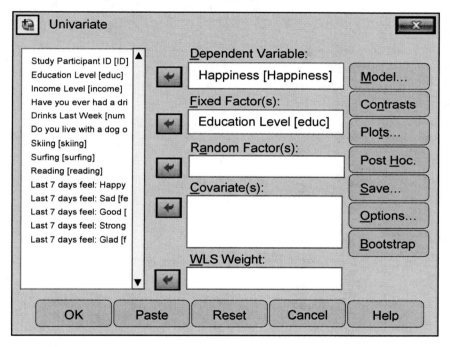

5) Click **Options.**

6) Within that dialogue box that opens, click **Estimates of Effect Size** and **Descriptive Statistics**, as presented in **Figure 4.97.**

7) Click **Continue.**

8) Click **OK** or **Paste.**

SPSS Output 4.27 presents the statistical output from the **General Linear Model**. The final column in the box labeled **Tests of Between-Subjects Effects** presents the *Partial Eta Squared (PES)*. In this column, we see that for the variable *educ* (*Education Level*) the effect size estimate of the PES is .349. Recall, the effect sizes for the partial Eta squared are small (>0.01), medium (>0.06), and large (>0.14). Thus, the predictor *Education Level* has a very large effect size (.349) on the dependent variable *Happiness*.

Figure 4.97 Activating **Estimates of Effect Size** and **Descriptive Statistics**

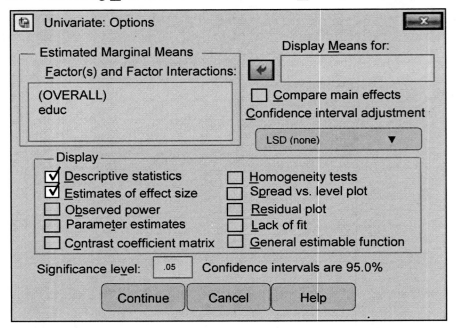

SPSS Output 4.27

Tests of Between-Subjects Effects

Dependent Variable: Happiness

Source	Type III Sum of Squares	df	Mean Square	F	Sig.	Partial Eta Squared
Corrected Model	16.69	2	9.85	25.96	.000	.349
Intercept	716.44	1	716	1888	.000	.951
Educ	19.69	1	9.85	25.96	.000	.349
Error	36.79	97	.38			
Total	809.96	100				
Corrected Total	56.484	99				

(3rd) Magnitude:

Large effect size (PES = .349)

Summary. The One-Way ANOVA indicated a statistically significant relationship between variables as the probability level for this test was less than .05 (*p*<.05). Actually, probability is reported as less than .001 (*p*<.001) as the level of probability was .000 (Sig.=.000). In terms of directionality, the mean scores, as well as the post hoc test indicated that a higher *Education Level* (Graduate School: *M*=3.23, *SD*=.55), was associated with a higher level of *Happiness*, relative to a lower Education Level (College: *M*=2.70, *SD*=.61; High School/GED: *M*=2.17, *SD*=.70). Lastly, in terms of magnitude, the *partial Eta squared* effect size (.35) indicated that covariate variable *Education Level* evidenced large effect upon the dependent variable *Happiness*.

4.6.2 Correlation

The Bivariate Test Key indicates a correlation is the appropriate statistical procedure examining the relationship between a continuous variable with another continuous variable. Therefore, we will use this procedure to examine the relationship between the continuous covariate variable *Income Level* (Response Range: $100k-$500k) with the continuous dependent variable *Happiness* (Response Range: 1-5). This procedure will indicate if *Happiness* scores either increase or decrease at a statistically significant level as *Income Levels* increase or decrease. Of the several types of tests of correlation, we will perform the **Pearson product-moment correlation** (the most commonly used). If a small sample size is an issue, we would consider a **Spearman's rank *non-parametric* correlation**, which is also available in SPSS. The Pearson's correlation has several assumptions, including **linearity** (the relationship between variables must be linear), **normality** (the distribution of scores for both variables must be approximately normal), and **no outliers** (the distributions should not contain significant outliers scores). These assumptions were examined in **Step 3: Checks of Data Integrity**.

To conduct the Pearson's correlation in SPSS go to:

1) **Analyze→ Correlation→ Bivariate**.

2) Then move the two continuous variables, *Income Level* and *Happiness*, from the left into the right hand column as presented in **Figure 4.98**.

3) Check the box next the term **Pearson**.

4) Click either **OK** or **Paste**.

Interpreting the Statistical Output: Pearson's Correlation. **SPSS Output 4.28** presents the statistical output for the **Pearson's correlation** procedure. In this section we use this statistical output to identify the three dimensions of the relationship between the dependent variable *Happiness* and the covariate variable *Income Level*. Specifically, we will present the relationship dimensions of significance, directionality, and magnitude in three separate parts (Part A, Part B, and Part C, respectively).

Figure 4.98 Pearson's correlation: *Income Level* and *Happiness*

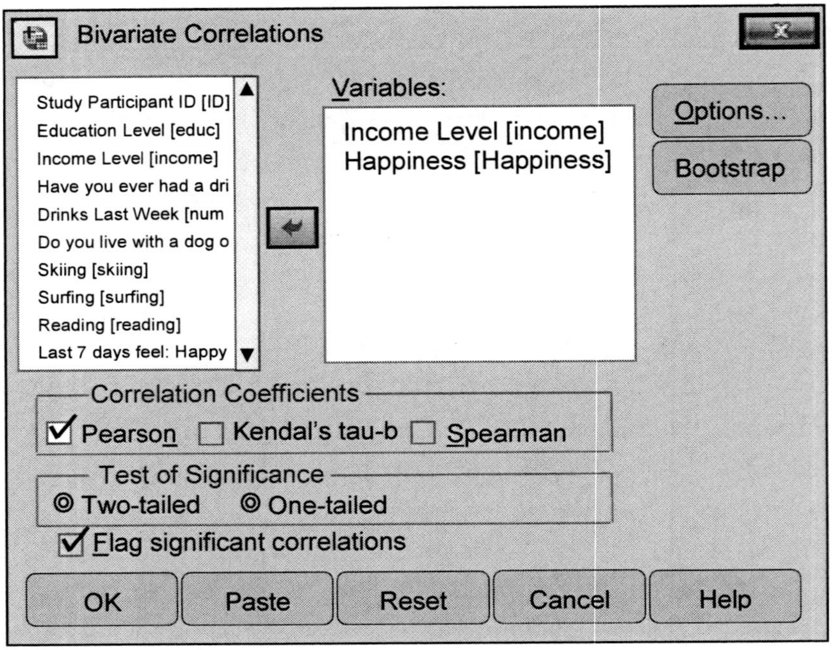

Part A Relationship Dimension 1: Significance

Within the single box produced in the statistical output, labeled **Correlations,** the same results are presented in each intersecting cell, so we could refer to either. We will use the cell at the end of the first row that begins with the variable name *Income Level*. Within this row, first refer to the second line labeled **Sig. (2-tailed)**. This line presents the **statistical significance** (probability level) for the procedure, which is .000 (in the far right hand column), indicating that the test is statistically significant ($p<.05$). Therefore, we will continue to examine the results of the test.

Part B Relationship Dimension 2: Directionality

Next, refer to the top row labeled *Pearson Correlation*, which is known as the Pearson *r* correlation coefficient, which will allow us to assess **directionality**. If the correlation coefficient begins with a minus sign, the relationship between the variables is negative. However, there is not a minus sign, which indicates a positive relationship whereas scores reflecting the variable *Income Level* increases, scores within the variable *Happiness* increase.

Part C Relationship Dimension 3: Magnitude

The Pearson r correlation coefficient also reflects the **magnitude** or "**effect size.**" Specifically, the Pearson r correlation coefficient can range -1 to +1. A value of 0.0

indicates no correlation between variables, but the closer the value is to 1.0 (- or +) the stronger the correlation. The correlation coefficient can also be squared to produce an **effect size** (reflecting the variance in one variable explained by the other variable). If the coefficient .884 is squared the resulting value is .7815, indicating that over 78% (78.15%) of the variance in the dependent variable *Happiness* is explained by the covariate variable *Income Level*. This indicates a large effect size as guidelines for the social sciences suggest a small, medium, and large effect size for the r^2 to be 0.10, 0.30, and 0.50 (r^2 to be 0.01, 0.09, 0.25), respectively (Cohen, 1992).

SPSS Output 4.28

Correlations

Correlations

		Income Level	Happiness
Income Level	Pearson Correlation	1	.884**
	Sig. (2-tailed)		.000
	N	100	100
Happ-iness	Pearson Correlation	.884**	1
	Sig. (2-tailed)	.000	
	N	100	100

**Significant at the 0.01 level (2-tailed).

(2nd) Directionality:
Positive relationship (no -)

(3rd) Magnitude:
Large effect size (r^2 = .78)

(1st) Significance:
Statistically significant

Summary. Overall, the Pearson's correlation indicated a statistically significant relationship between variables as the probability level for this test was less than .05 ($p<.05$). Actually, probability is reported as less than .001 ($p<.001$) as the level of probability was .000 (Sig.=.000). In terms of directionality, the correlation coefficient was positive, which indicates a positive relationship where *Happiness* scores increase as *Income Level* increases. Lastly, in terms of magnitude, after squaring the correlation coefficient (r = .78 effect size) we saw that the covariate variable *Income Level* evidenced a large effect upon the dependent variable *Happiness*.

4.6.3   Independent-Samples T-Test

The bivariate test key indicates the **independent-samples t-test** is the appropriate statistical procedure to use when examining a continuous variable with a categorical predictor with two response categories. Therefore, we will use this procedure to examine the relationship between the categorical independent variable *Do You Live with a Dog or Cat?* with the continuous dependent variable *Happiness*. This procedure will indicate if there is a statistically significant difference between mean *Happiness* scores (Response Range: 1-5), among study participants that live with a dog or cat.

The independent samples t-test procedure has certain assumptions, including **independence of observations** (there is no relationship between the study participants in each group or between the groups themselves, such as there are different participants in each group with no participant being in more than one group), **no significant outliers** (the distributions of scores in the continuous variable should not contain significant outliers scores), **normality** (the scores within the continuous variable are normally distributed), and **homoscedasticity** (homogeneity of variances). These assumptions were examined in **Step 3: Checks of Data Integrity**.

Again, as in the One-Way ANOVA, it is important to consider the number of study participants within each response category within the categorical variable employed in the statistical procedure as this relates to statistical power (Kraemer & Thiemann, 1987). Here, a minimum of *30 participants per cell* has been suggested to detect a medium-to-large effect size (Cohen, 1988). If small numbers of study participants per cell presents an issue, you may want to consider a **Mann-Whitney U Test** *non-parametric Independent-Samples test* **equivalent**, which is also available in SPSS. To conduct the independent Samples t-test in SPSS go to:

1) **Analyze→ Compare Means→ Independent Samples T-Test**.

2) Move the continuous variable (*Happiness*) from the left hand column into the cell on the right labeled **Test Variable(s)**.

3) Move the categorical variable (*Do You Live with a Dog or a Cat?*) from the left hand column into the cell on the right labeled **Grouping Variable**.

4) Click **Define Groups**.

5) In the dialogue box that opens enter the value for the first response category, such as a '1' for *dog* in the **Group 1** cell, then value for the second response category such as '2' for *cat* in the cell for **Group 2** (see **Figure 4.99**).

6) Click **Continue**.

7) The original dialogue box will look similar to **Figure 4.100**.

8) Click **OK** or **Paste**.

Figure 4.99 Define groups: *Do You Live with a Dog or a Cat?*

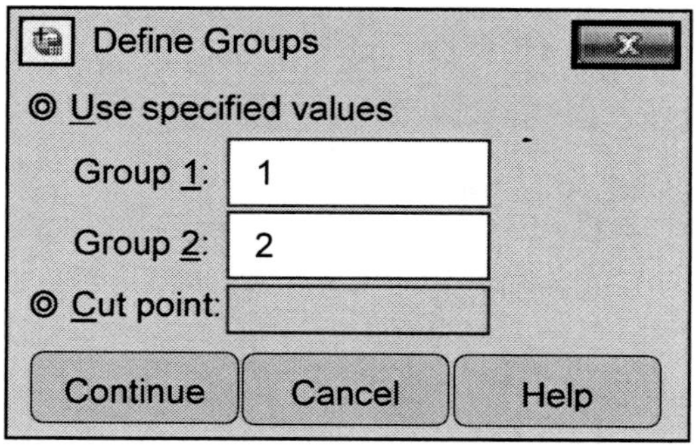

Figure 4.100 Independent-samples t-test: *Happiness* by *Live with a Dog or a Cat?*

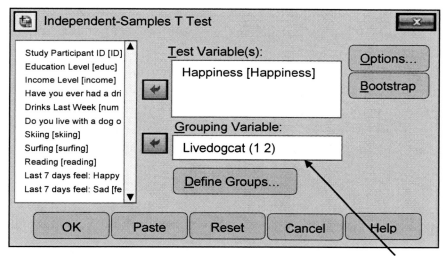

Do You Live with a Dog (=1) or a Cat (=2)?

Interpreting the Statistical Output: Independent-Samples T-Test. **SPSS Output 4.29** presents the statistical output for the independent-samples t-test procedure. In this section we use this statistical output to identify the three dimensions of the relationship between the dependent variable *Happiness* and the independent variable *Do You Live with a Dog or a Cat?* Specifically, we will present the relationship dimensions of significance, directionality, and magnitude in three separate parts (Part A, Part B, and Part C, respectively).

Part A Relationship Dimension 1: Significance

Within the statistical output, the second box labeled **Independent-Samples T-Test** presents the level of statistical significance for the procedure. First refer to the column to the left labeled **Levene's Test for Equality of Variances**. The subcolumn beneath labeled **Sig.**, will indicate if the Levene's Test is statistically significant ($p<.05$). If the Leven's test is statistically significant it indicates that the variances in scores reflecting the continuous variable (*Happiness*) are significantly unequal among the two categorical response groups (lives with dog, lives with cat). Specifically, if the value in the subcolumn Sig. is .05 ($p<.05$) or below, the values in the second row labeled **Equal Variances not assumed** should be used. In this analysis Sig.=.872, indicating the Levene's test is not statistically significant ($p>.05$) and the top row of values (**Equal Variances assumed**) should be used. On this line under the column labeled **Sig. (2-tail) is .000** (which reflects the level of statistical significance for the t-test), **indicating that the difference in mean scores is statistically significant ($p<.05$).**

Part B   Relationship Dimension 2: Directionality

Top box in the output (labeled **Group Statistics**) presents the mean *Happiness* scores for the two categorical groups included in the t-test procedure (study participants that live with a *dog* or *cat*). Within the column labeled *Do You Live with a Dog or Cat?* the first row labeled *Dog* reflects there are 47 study participants that live with a dog ($n=47$) with a mean *Happiness* score of 3.02 ($SD=.79$). The second row labeled *Cat* reflects there are 53 study participants that live with a cat ($n=53$) with a mean *Happiness* score of 2.48 ($SD=.62$). Thus, the independent-samples t-test indicates that higher ***Happiness* scores are associated with being in the group living with a dog relative to a cat.**

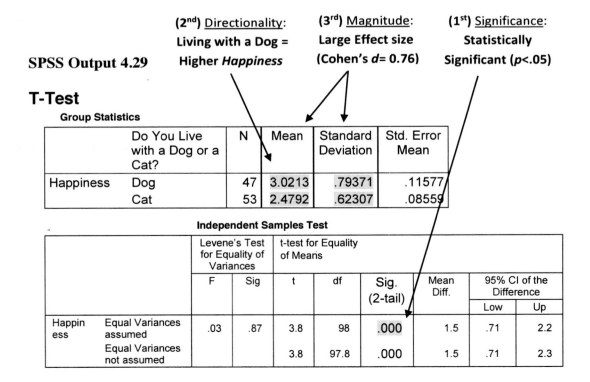

SPSS Output 4.29

(2ⁿᵈ) Directionality: **Living with a Dog = Higher *Happiness***

(3ʳᵈ) Magnitude: **Large Effect size (Cohen's *d*= 0.76)**

(1ˢᵗ) Significance: **Statistically Significant (*p*<.05)**

T-Test

Group Statistics

	Do You Live with a Dog or a Cat?	N	Mean	Standard Deviation	Std. Error Mean
Happiness	Dog	47	3.0213	.79371	.11577
	Cat	53	2.4792	.62307	.08559

Independent Samples Test

		Levene's Test for Equality of Variances		t-test for Equality of Means					95% CI of the Difference	
		F	Sig	t	df	Sig. (2-tail)	Mean Diff.		Low	Up
Happiness	Equal Variances assumed	.03	.87	3.8	98	.000	1.5		.71	2.2
	Equal Variances not assumed			3.8	97.8	.000	1.5		.71	2.3

Part C   Relationship Dimension 3: Magnitude

A common measure of effect size when using an independent-samples t-test is the Cohen's d. While the Cohen's d is not computed via the independent-samples t-test in SPSS, computing this value is quite straightforward (the formula is presented in **Figure 4.101**). Specifically, to compute Cohen's d:

1) **Compute the denominator (the *pooled standard deviation* presented in Figure 4.102):**

   a. Square the **standard deviation** for each group:
      SD of Lives with a Dog: .79371 = .6299755641
      SD of Lives with a Cat: .62307 = .3882162249
   b. Sum those values: .6299755641 + .3882162249 = 1.018191789
   c. Divide that summed number by 2: 1.018191789/2 = .5090958945
   d. Take the square root of that number: .71350956160376715

2) **Compute the numerator**: Subtract the mean score of group 1 (3.0213) from the mean of group 2 (2.4792), which is = 0.5421.

3) **Divide the numerator by the denominator**: 0.5421/.71350956160376715 = 0.7598.

Figure 4.101   Formula to compute Cohen's d

$$d = \frac{Mgroup1 - Mgroup2}{SDpooled}$$

Figure 4.102   The *pooled standard deviation* to compute the Cohen's d

$$SDpooled = \sqrt{(SD^2group1 + SD^2group2)/2}$$

An easier means of computing the Cohen's d is to use one of the many calculators available free online (e.g., http://www.uccs.edu/~lbecker/), which will compute this formula for you. For example, **Figure 4.103** presents a free online calculator with the mean and standard deviation values from our independent-samples t-test entered for **Group 1** and **Group 2**. Once the **Compute** button has been clicked the Cohen's d effect size is produced, which for our analysis is 0.759765 (which matches our manual computation). Recall, the small, medium, and large Cohen's d effect sizes are 0.2, 0.5, and 0.8, respectively. Thus, our d of 0.7598 reflects an approximately large effect size between the two variables. Specifically, we would say that the independent variable *Do You Live with a Dog or a Cat?* has a large effect on the dependent variable *Happiness*.

Figure 4.103   Free online calculator used to compute the Cohen's *d*

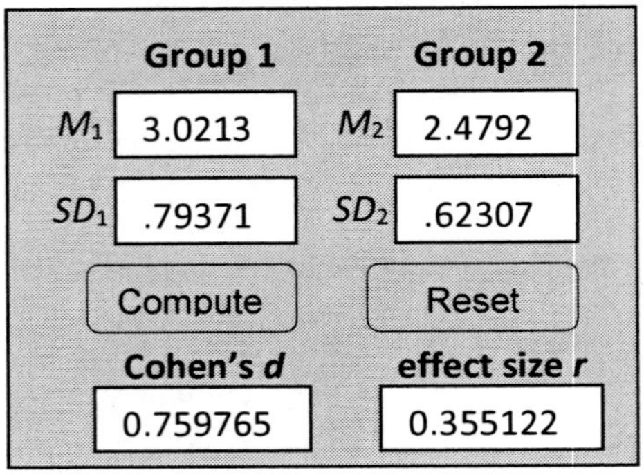

	Group 1		Group 2
M_1	3.0213	M_2	2.4792
SD_1	.79371	SD_2	.62307
	Compute		Reset
	Cohen's *d*		**effect size *r***
	0.759765		0.355122

Summary. Overall, the independent-samples t-test indicated a statistically significant relationship between variables (i.e., a difference between mean scores) as the probability level for this test was less than .05 ($p<.05$). Actually, probability is reported as less than .001 ($p<.001$) as the level of probability was .000 (Sig.=.000). In terms of directionality, a higher level of *Happiness* was associated with living with a dog (M=3.02, SD=.79) relative to a cat (M=2.48, SD=.62). Lastly, in terms of magnitude, the Cohen's *d* reflected that the independent variable *Do You Live with a Dog or a Cat?* evidenced a large effect (Cohen's *d* effect size=0.7598) upon the dependent variable *Happiness*.

4.6.4   What Did Bivariate Analysis Tell Us?

Within the context of sample study one, our primary goal in bivariate analysis is to identify which predictor variables (covariate/independent variables) have a statistically significant relationship with the dependent variable so these predictors may be included in **Step 6: Multivariate Analysis**. Subsequently, although we examined all three dimensions of the relationships between variables (significance, directionality, and magnitude) in our bivariate testing, in this section we will only review the dimension of statistical significance.

Figure 4.104 reflects the results on the three bivariate tests used in sample study one examining which predictor variables have a statistically significant association with the continuous dependent variable *Happiness* (1-5). Notice that we have entered an arrow connecting covariate variable 1: *Education Level* with the dependent variable, which indicates a One-Way ANOVA revealed a statistically significant relationship ($p<.05$). Likewise, there are arrows reflecting statistically significant relationships between covariate variable 2: *Income Level* and the independent variable: *Do You Live with a Dog or a Cat?* with the dependent variable.

Figure 4.104   What Bivariate Analysis Tells Us for Sample Study One?

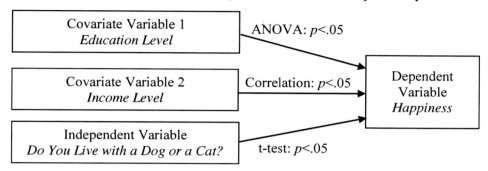

Subsequently, the findings produced through bivariate analysis tells us that all three predictors are significantly related to the dependent variable, and therefore will be included in the next step of multivariate analysis. Please also note, at this step we can only see which predictors are significantly related to the dependent variable, but we cannot see which variable is the strongest predictor of *Happiness* (1-5). That dimension will be revealed in multivariate analysis.

4.7   STEP 6: MULTIVARIATE ANALYSIS

If we break down the terms within *multivariate analysis* we see the roots of the term suggest examining the relationship between multiple (*multi*) variables (*var*). This is appropriate as we will examine the relationship between *multiple* predictor variables with the dependent variable. Specifically, recall in the previous step we employed bivariate analysis to identify which predictor variables (covariate and independent variables) are related to the dependent variable at a statistically significant level ($p<.05$). In this step we will employ multivariate analysis to identify which predictor variable (*Education Level, Income Level, Do You Live with a Dog or a Cat?*) is the strongest predictor of the dependent variable *Happiness*, which was the overall goal of our data analysis study.

The Multivariate Test Key presented in **Figure 4.105**, describes that selecting the appropriate multivariate statistical procedure depends upon the structure (i.e., categorical or continuous) of the dependent variable. **Figure 4.106** presents the **Multivariate Test Key with the study variable plugged in**. Specifically, we see the key with the dependent variable for sample study 1 plugged in. The key instructs that the appropriate regression model to use when examining a continuous dependent variable, such as *Happiness* (Response Category: 1-5) is a multiple linear regression. **When linear regression involves more than one predictor, the procedure is termed** *multiple linear regression*, **which is the term we use for this procedure.**

Multiple linear regression can be used to predict the value of the continuous dependent variable based on the value of two or more other predictor variables. Multiple linear regression will also allow us to determine the overall fit (variance explained) of the model and the relative contribution of each of the predictors to the total variance explained in the dependent variable. We know these data are appropriate for this procedure as in **Step 3: Checks of Data Integrity**, we assessed if several assumptions of multiple linear regression analysis were met including **linearity, normality, no undue influence of outlier scores, independence of observations, homoscedasticity,** as well as other necessary considerations.

Figure 4.105   The Multivariate Test Key

Dependent Variable	Regression to Use
Categorical (≥3 categories)	Multinomial Logistic (Not Presented here)
Categorical (2 categories)	Binary Logistic
Continuous	Linear

Figure 4.106   Multivariate Test Key with the study variable plugged in

Dependent Variable	Regression to Use
Categorical (≥3 categories)	Multinomial Logistic (Not Presented here)
Categorical (2 categories)	Binary Logistic
Continuous	Linear *(>1 Predictor = Multiple Linear)*
Continuous (1-5)	Happiness (1-5)

4.7.1   Dummy-Coding Variables

Dummy-Coding is the key to using a categorical variable with three or more response categories in a regression model. Before the analysis is performed, an important consideration must be made if a categorical predictor variable with three or more response categories is to be included in the regression model. Specifically, if such a categorical variable will be used in the regression model, each response category within the original categorical predictor variable must be **dummy-coded** into a separate **Yes** or **No** dichotomous categorical variable.

For example, in the regression model for sample study one, the categorical predictor variable *Education Level* will be included, which has three response categories (High School/GED, College, and Graduate School). To use the variable *Education Level* in the regression model, each response category must be made into a separate Yes or No dichotomous categorical variable, which we will describe below. First, consider the original single variable *Education Level* along with the three response categories (note in SPSS data base response categories are called *Values*). Recall, when we entered these data into the SPSS database, we ascribed the categories the numbers 1, 2, and 3, as presented below:

Response Category (Value) 1: High School/GED   (Coded as "1" in SPSS)
Response Category (Value) 2: College   (Coded as "2" in SPSS)
Response Category (Value) 3: Graduate School   (Coded as "3" in SPSS)

When we dummy-code the variable *Education Level*, we create **three new Yes/No variables** to use in the regression model in place of the original single variable. We will now create these new variables in SPSS using the **Transform** function.

Creating Dummy Coded Variable #1: EducationHSGED. First, we will dummy-code to produce Variable 1 *EducationHSGED*, to identify study participants with a highest level of education at the High School/GED level. To do this in SPSS, go to:

1) **Transform→ Recode Into Different Variables→**.

2) In the dialogue box that opens, move the variable *Education Level* from the left hand column into the right hand column using the center arrow.

3) Under the box labeled **Name**, enter the name of the first new dummy-coded variable, *EducationHSGED*.

4) Click **Change**, which will bring the new variable name into the center column.

5) In the box marked **Label**, enter a description of the new variable (e.g., Education Level: High School/GED) (**Figure 4.107** presents the dialogue box).

6) Click the box labeled **Old and New Values**.

7) In the dialogue box that opens, click **Value** in upper left hand column (under **Old Value**) and enter a '1' in the cell (1=HS/GED within *Education Level*).

8) Click **Value** in upper right hand column (under **New Value**) and enter a '1' (1=Yes) in the cell, click **Add**.

9) Click **Value** in upper left hand column (under **Old Value**) and enter a '2' in the cell (2=College within *Education Level*).

Figure 4.107 Dummy-coding the category *High School/GED*

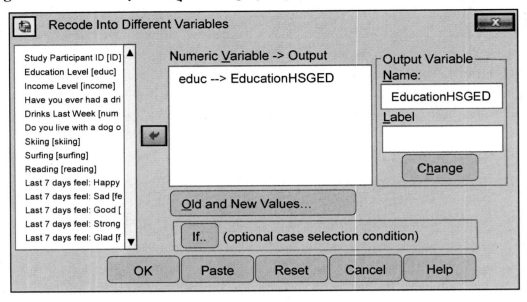

Figure 4.108 Coding for the dummy-coded category *High School/GED*

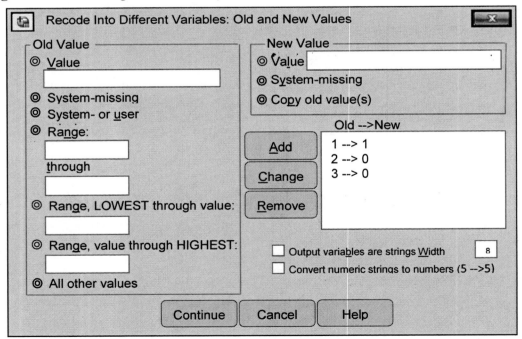

10) Click **Value** in upper right hand column (under **New Value**) and enter a '0' (0=No) in the cell, click **Add**.

11) Click **Value** in upper left hand column (under **Old Value**) and enter a '3' in the cell (3=Graduate School within *Education Level*).

12) Click **Value** in upper right hand column (under **New Value**) and enter a '0' (0=No) in the cell, click **Add** (**Figure 4.108** presents the dialogue box).

13) Click **Continue**

14) Click **OK** or **Paste**.

Creating Dummy Coded Variable #2: EducationCollege. Next, we will dummy code to produce Variable 2 *EducationCollege*, which will indicate the study participant with a highest level of education at the College level. To do this in SPSS go to:

1) **Transform→ Recode Into Different Variables**.

2) In the dialogue box that opens, move the variable *Education Level* from the left hand column into the right hand column using the center arrow.

3) Under the box labeled **Name**, enter the name of the first new dummy-coded variable, *EducationCollege*.

4) Click **Change**, which will bring the new variable name into the center column.

5) In the box marked **Label**, enter a description of the new variable (e.g., Education Level: College) (presented in **Figure 4.109**).

Figure 4.109 Dummy-coding the category *College*

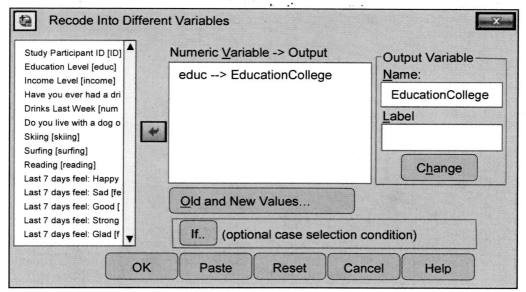

6) Click the box labeled **Old and New Values**.

7) In the dialogue box that opens, click **Value** in upper left hand column (under **Old Value**) and enter a '1' in the cell (1=HS/GED within *Education Level*).

8) Click **Value** in upper right hand column (under **New Value**) and enter a '0' (0=No) in the cell, click **Add**.

9) Click **Value** in upper left hand column (under **Old Value**) and enter a '2' in the cell (2=College within *Education Level*).

10) Click **Value** in upper right hand column (under **New Value**) and enter a '1' (1=Yes) in the cell, click **Add**.

11) Click **Value** in upper left hand column (under **Old Value**) and enter a '3' in the cell (3=Graduate School within *Education Level*).

12) Click **Value** in upper right hand column (under **New Value**) and enter a '0' (0=No) in the cell, click **Add** (presented in **Figure 4.110**).

13) Click **Continue**.

14) Click **OK** or **Paste**.

Figure 4.110   Coding for the dummy-coded category *College*

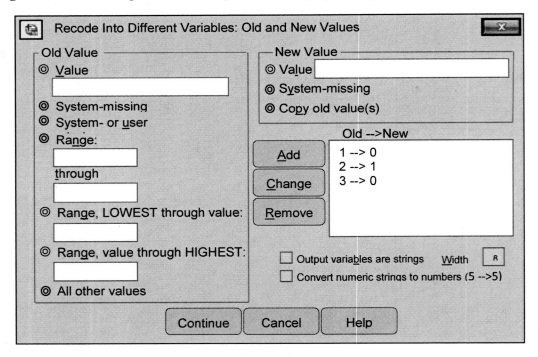

Creating Dummy Coded Variable #3: EducationGraduateSchool. Next, we will dummy code to produce Variable 3 *EducationGraduateSchool*, which is the variable that will indicate the study participants that have a highest level of education at the Graduate School level. To do this in SPSS, go to:

1) **Transform→ Recode Into Different Variables**.

2) In the dialogue box that opens, move the variable *Education Level* from the left hand column to the right hand column using the center arrow.

3) Under the box labeled **Name**, enter the name of the first new dummy-coded variable, *EducationGraduateSchool*.

4) Click **Change**, which will bring the new variable name into the center column.

5) In the box marked **Label**, enter a description of the new variable (e.g., Education Level: Graduate School) presented in **Figure 4.111**).

Figure 4.111   Dummy-coding the category *Graduate School*

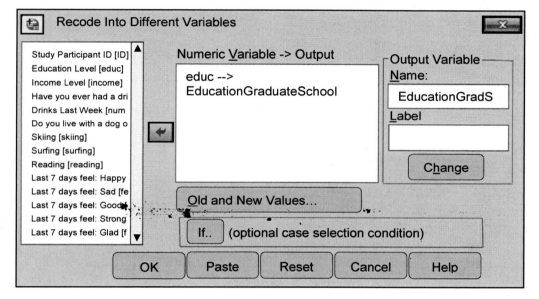

6) Click the box labeled **Old and New Values**.

7) In the dialogue box that opens, click **Value** in upper left hand column (under **Old Value**) and enter a '1' in the cell (1=HS/GED within *Education Level*).

8) Click **Value** in upper right hand column (under **New Value**) and enter a '0' (0=No) in the cell, click **Add**.

9) Click **Value** in upper left hand column (under **Old Value**) and enter a '2' in the cell (2=College within *Education Level*).

10) Click **Value** in upper right hand column (under **New Value**) and enter a '0' (0=No) in the cell, click **Add**.

11) Click **Value** in upper left hand column (under **Old Value**) and enter a '3' in the cell (3=Graduate School within *Education Level*).

12) Click **Value** in upper right hand column (under **New Value**) and enter a '1' (1=Yes) in the cell, click **Add** (presented in **Figure 4.112**).

13) Click **Continue**.

14) Click **OK** or **Paste**.

Figure 4.112 Coding for the dummy-coded category *Graduate School*

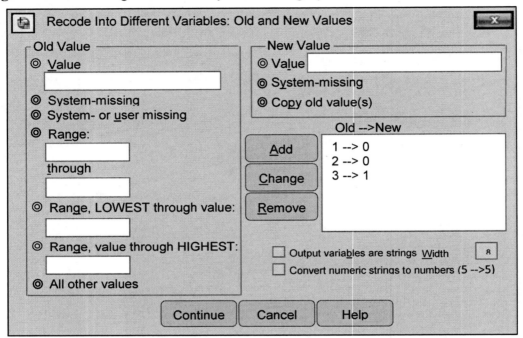

***What the Dummy Coded Variables Look Like in the SPSS Database.* Figure 4.113** presents the **Data View** portion of the SPSS database, describing the three new dummy coded variables that correspond to the original variable *Education*. For example, the first dummy coded variable *EducationHSGED*, indicates the number 1 (=yes) when the original variable *Education* is 1 (=HS/GED). Otherwise there is a number 0 (=No). This same pattern is evident for the other values (*College, Graduate School*) within the original *Education* variable relative to the new dummy coded variables.

Please note the *values label* icon ![icon]. Instead of pretending numbers are words, such as a *1* is a *Yes* and a *0* is a *No*, you can click this icon and it will show the corresponding world to the numbers in the SPSS database. For example, if you click the *values icon*, **Figure 4.113** will become the image in **Figure 4.114**. It may be easier to examine the cells in this form.

Figure 4.113 The dummy-coded variable *EducationHSGED* in **Data View**

Education	EducationHSGED	EducationCollege	EducationGraduateSchool
1.00	1.00	0.00	0.00
1.00	1.00	0.00	0.00
2.00	0.00	1.00	0.00
2.00	0.00	1.00	0.00
3.00	0.00	0.00	1.00
3.00	0.00	0.00	1.00

1= HS/GED in the original
variable *Education Level*

1=Yes in the new dummy coded
variable *EducationHSGED*

Figure 4.114 The dummy-coded variable *EducationHSGED* in text response

Education	EducationHSGED	EducationCollege	EducationGraduateSchool
HS/GED	Yes	No	No
HS/GED	Yes	No	No
College	No	Yes	No
College	No	Yes	No
Graduate	No	No	Yes
Graduate	No	No	Yes

After Dummy Coding You Must Select a Reference Group. Once the values (e.g., *High School/GED, College, and Graduate School*) within the original categorical variable (e.g., *Education Level*) have been dummy coded, you must specify which value will be used as the **Reference Group** in the multiple linear regression model.

The **Reference Group** is the group to whom the other dependent variable scores will be compared. For example, if the group *High School/GED* within the variable *Education Level* is made the **Reference Group** in our study of levels of *Happiness* (Dependent Variable), then the analysis will tell us: Do *Happiness* Scores for those with a *College* and/or *Graduate school* level of education differ significantly **in reference to** those with a *High School/GED* level of education? Thus, essentially, the **reference group** is the group the other groups are **compared to (or referenced)** regarding the scores on the dependent variable.

Selecting the Reference Group. There are several reasons for selecting one group as the reference group instead of the others. The first consideration is making the group of interest or under study the reference group. For example, if our study is focused on the well-being of individuals that have a highest level of education of *High School/GED*, then it would make sense to make that group the reference group. Then we would see the scores of *High School/GED, College and Graduate School* relative to the group of interest *High School/GED*. The dummy-coded variable that represents the reference group (e.g., *EducationHighSchoolGED*) is left out of the multiple linear regression model, which will indicate to the SPSS software that this is the reference group.

4.7.2 Conducting Multiple Linear Regression Analysis

To conduct the multiple linear regression analysis procedure in SPSS go to:

1) **Analyze→ Regression→ Linear.**

2) In the dialogue box that opens, move the dependent variable *Happiness* from the left hand column into the box labeled **Dependent** on the right.

Figure 4.115   Multiple linear regression analysis: Predicting *Happiness*

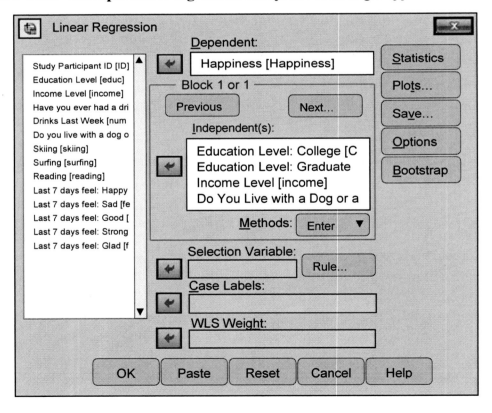

3) Enter the predictor variables from the left hand column into the right hand column, labeled **Independent(s),** including: a) the dummy-coded covariate variables *Education College* and *Education Graduate School* (leave out

reference group *Education High SchoolGED*); The covariate variable *Income Level*; and the independent variable *Do You Live with a Dog or a Cat?*(as presented in **Figure 4.115**).

4) Click **OK** or **Paste**.

For the sake of keeping things simple, we did not click the diagnostics described earlier regarding checks of test assumptions available in the linear regression function toward assessing linearity, homoscedasticity, independence of errors, and multicollinearity. However, typically these may also be included here.

Interpreting the Statistical Output: Multiple Linear Regression Analysis. **SPSS Output 4.30** presents the statistical output for the **Multiple Linear Regression Analysis** procedure. In this section we use this statistical output to identify the three dimensions of the relationship between the dependent variable *Happiness* and predictor variables (covariate and independent variables) in the context of the full model. Specifically, we will present the relationship dimensions of significance, directionality, and magnitude in three separate parts (Part A, Part B, and Part C, respectively).

Part A   Relationship Dimension 1: Significance

Overall Model. Note the second box in the output labeled **ANOVA**. In the final column labeled **Sig.**, the level of probability (statistical significance) for the overall test is indicated, which is **.000**. Thus, the overall model for the **multiple linear regression model** procedure is statistically significant ($p<.05$). Therefore, we will continue to examine the results of the test.

Individual Predictor Variables. Next, refer to the final table labeled **Coefficients**. This table provides information regarding the relationship between each predictor variable with the dependent variable. First, refer to the final column labeled **Sig.**, which presents the probability levels for these relationships. We see that all values are below .05 ($p<.05$), which indicates that all the predictor variables have a statistically significant relationship with the dependent variable. If a value greater that .05 ($p>.05$) we would not consider the relationship between that predictor and the dependent variable.

Part B   Relationship Dimension 2: Directionality

The Unstandardized Beta (regression coefficient). Within the **SPSS Output 4.30**, refer to the final table labeled **Coefficients**. Refer to the first column under the label Unstandardized Coefficients labeled **B**, which indicates the directionality of the relationships between the predictor variables with the dependent variable. Specifically, this column presents the Unstandardized Beta. For example, observe the two dummy-coded variables at the top of the first column, *Education Level: College* and *Education Level Graduate School*. Notice that the unstandardized beta for each of these variables does not begin with a minus sign (.382 and .698) indicating a positive relationship. However, recall there is a Reference group (*Education Level: High School/GED*) for these groups.

Therefore, the positive betas, along with the values reflecting statistical significance ($p<.05$ as Sig. = .000 and .000), indicate that study participants with a level of education at the *College* and *Graduate School* levels have significantly higher scores (recall positive relationship indicates higher scores) reflecting *Happiness* **in reference to** study participants with a *High School/GED* level of education.

Next, refer to covariate predictor variable *Income Level*. Again, we see under the Unstandardized Coefficients column, that the value for the beta (**B**) is positive (0.01) as there is not a minus sign, indicating a positive association. Specifically, this positive unstandardized beta indicates that as the covariate predictor variable *Income Level* increases, the dependent variable *Happiness* also increases. However, notice that the beta (**B**) for the independent variable *Do You Live with a Dog or Cat?* is negative as there is a minus sign in front of the unstandardized beta.

In order to interpret the meaning of the negative unstandardized beta, recall that living with a dog is coded as 1 and a cat 2 in the SPSS database. Thus, the negative beta suggests as the predictor *Do You Live with a Dog or Cat?* increases from 1 to 2, or in conceptual terms from a *dog* to a *cat*, *Happiness* scores go down (the minus indicates a negative relationship where scores decrease). In other words, living with a cat is associated with significant decreases in *Happiness* scores.

More About the Unstandardized Beta. It is important to note that the unstandardized beta also serves another purpose beyond indicating the directionality of a relationship between variables. Specifically, **the unstandardized beta also reflects the degree of change in the dependent variable relative to changes in the predictor variable**. Although a full discussion of the mechanics of the model in this respect is a bit beyond the focus of this text, this is an important consideration and fact to know. As a brief presentation, consider the classic model equation expressing this modeled change presented in **Figure 4.116**. In this model:

Y represents the **dependent variable**

X reflects a **predictor variable** (**independent** or **covariate variable**) response categories

b reflects the actual value of the **unstandardized beta** in the statistical output

Figure 4.116 The classic multiple linear regression model function equation

$$Y = b_0 + b_1X_1 + b_2X_2 + b_3X_3$$

Thus, because we know our dependent variable, predictor variables (and response categories), and the unstandardized betas (as per the SPSS output), we can plug these numbers in and fill in the model in **Figure 4.117**. However, to keep these simple, let's just examine the *rate of change between one predictor variable, the independent variable (Do You Live with a Dog or a Cat?) and the dependent variable (Happiness).* This equation is presented in **Figure 4.117**, where we see:

1) The first arrow indicates the **Y** represents the **dependent variable** *Happiness.*

2) The second arrow indicates the **b** reflects the value of **unstandardized beta** as per the SPSS statistical output (which is **-0.461**) for the independent variable *Do You Live with a Dog or a Cat?*

3) Finally, the third arrow indicates the **X** represents the coding of the response categories within the **independent variable, which are 1 and 2,** where for the independent variable *Do You Live with a Dog or a Cat?*, Dog=1 and Cat=2.

Figure 4.117 Linear regression model function equation with only one predictor

$$Y \quad = \quad b_1X_1$$

Happiness = (-0.461) (X = '1' or '2')

The change in the dependent variable occurs according to the response given for the independent variable. For example, if a study participant indicates he or she **lives with a cat** (coded with a **value of 2**) then the equation becomes: *Happiness* = (-0.461) (2).

Therefore, for study participants that live with a cat, their *Happiness* scores decreases by .922 on average as: *Happiness* = (-0.461) (2), which is -.922.

If a study participant indicates he or she lives with a dog (coded with a value of 1) then the equation becomes: *Happiness* = (-0.461) (1).

Therefore, for study participants that live with a dog, their *Happiness* score decreases by .461 on average as: *Happiness* = (-0.461) (1), which is -.461.

SPSS Output 4.30

Regression

(3ʳᵈ) <u>Magnitude</u>:
Large R^2 effect size

Model Summary[b]

Model	R	R Square	Adjusted R Square	Std. Error of the Estimate
1	.784	.615	.599	.47856

(1ˢᵗ) <u>Significance (model)</u>:
Statistically significant ($p<.05$)

ANOVA[a]

Model		Sum of Squares	df	Mean Square	F	Sig.
1	Regression	34.727	4	8.682	37.91	.000
	Residual	21.757	95	.229		
	Total	56.484	99			

(1ˢᵗ) <u>Significance (predictors)</u>:
Statistically significant ($p<.05$)

Coefficients[a]

Model		Unstandardized Coefficients		Standardized Coefficients	t	Sig.
		B	Std. Error	Beta (β)		
1	(Constant)	3.43	.183		18.7	.000
	Education Level: College	.382	.123	.252	3.7	.000
	Education Level: Graduate School	.698	.118	.333	5.6	.000
	Income Level	.010	.002	.386	5.7	.000
	Do You Live with a Dog or a Cat?	-.461	.097	-.306	-4.7	.000

(2ⁿᵈ) <u>Directionality</u>:
Unstandardized Beta (B) indicates all positive relationships except for the IV

(3ʳᵈ) <u>Magnitude</u>:
Standardized Beta (β) indicates the strongest predictor

Part C Relationship Dimension 3: Magnitude

Within the multiple linear regression model there are several means of indicating an effect size between the predictor and dependent variables. For example, the multiple linear regression analysis produces the standardized beta which indicates the strongest predictor within the regression model. In the regression model for sample study one, the highest standardized beta is .583 for the variable *Education Level: Graduate School*, which suggest this is the strongest predictor in the multiple linear regression model. However, the standardized beta is not the traditional effect size we might like to use to estimate the effect of the predictor variable on the dependent variable.

The multiple linear regression model also produces the R effect size statistic that reflects the amount of variance in the dependent variable explained by the predictor variable. In sample study one the R^2 is .615, which as per the **Figure 3.13** is a large effect size. It is also important note the **adjusted R^2**, which is a modified version of R statistic. Specifically, the adjusted R estimates how much variance in the dependent variable would be explained by the predictor variable(s) if the analysis was based upon the entire population from which the sample was derived. Generally, the closer the R^2 and **adjusted R^2**, the more generalizable and credible is the estimate of the R^2.

For example, the R^2 and **adjusted R^2**, in **SPSS Output 4.30** are quite similar (.615 and .599, respectively), which builds confidence in the generalizability of the findings. However, this effect size reflects the magnitude of the relationship between all the predictors and the dependent variable. Thus, we need another method to estimate the magnitude of the relationship between each individual predictor and the dependent variable.

However, the linear regression analysis procedure in SPSS does not have an option to estimate the effect size between variables. However, an equivalent procedure will produce the effect size estimate of the partial Eta squared statistic through the ANCOVA (Analysis of Covariance) statistical procedure. To perform an ANCOVA in SPSS go to:

1) **Analyze→ General Linear Models→ Univariate.**

2) Within the dialogue box that opens, move the dependent variable *Happiness* from the left hand column into the top cell under the term **Dependent Variable.**

3) Enter the categorical independent variable *Do You Live with a Dog or a Cat?* from the left hand column into the cell under the term **Fixed Factor(s).**

4) Move the covariate predictor variables *Education Level* and *Income Level* from the left hand column into the cell under the term **Covariate(s)** (as presented in **Figure 4.118**).

5) Realize here we are treating the covariate variable *Education Level* as a continuous covariate predictor variable in this model. This is done for several reasons, including the fact that within an ANCOVA model, the covariates should be continuous in nature. However, as noted earlier, it is often acceptable to treat ordinal variables (ranked categorical variables) as continuous variables.

6) Click **Options**.

7) Within the dialogue box that opens, click the box for **Estimates of Effect Size** and **Descriptive Statistics** (as presented in **Figure 4.97**).

8) Click **Continue**.

9) Click **OK** or **Paste**.

Figure 4.118   ANCOVA: Producing an effect size for predictors of *Happiness*

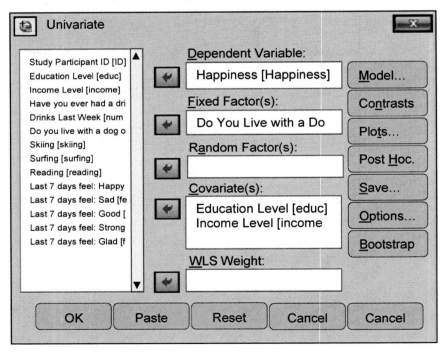

SPSS Output 4.31 presents statistical output for the ANCOVA procedure, which includes the partial Eta squared effect size. Specifically, the last column in the output labeled *Partial Eta Squared* presents the effect size estimates for the predictor/dependent variable relationships. Recall, the effect sizes for the partial Eta squared are small (>0.01), medium (>0.06), and large (>0.14). Thus, living with a dog or cat has a medium to large effect size (0.092), as does *Education Level* (0.111), while *Income Level* (0.646) has a very large effect size. These estimates reflect the strength of predictors described by the Standardized Beta in the multiple linear regression model where *Income Level* was the strongest predictor of *Happiness*.

SPSS Output 4.31

Tests of Between-Subjects Effects
Dependent Variable: Happiness

Source	Type III Sum of Squares	df	Mean Sqr	F	Sig.	Partial Eta Squared
Corrected Model	46.100	3	15.37	142.06	.000	.816
Intercept	23.229	1	23.23	214.75	.000	.691
educ	1.293	1	1.29	11.95	.001	.111
income	18.923	1	18.92	174.94	.000	.646
livedogcat	1.057	1	1.06	9.78	.002	.092
Error	10.384	96	.11			
Total	803.960	100				
Corrected Total	56.484	99				

(3rd) Magnitude:
IV: Medium to Large effect size (PES = .092)

4.7.3   What Did Multivariate Analysis Tell Us?

Figure 4.119 reflects the results on the multiple linear regression analysis used in sample study one examining which predictor variable is the strongest predictor of the dependent variable *Happiness*. We saw that at the multivariate level, all predictors evidenced a statistically significant relationship ($p<.05$) with the dependent variable. After the effects of both covariate predictor variables *Education Level* and *Income Level* were controlled for in the multivariate model, the independent variable *Do You Live with a Dog or Cat?* was related to the dependent variable *Happiness* at a statistically significant level. Furthermore, analysis indicated that study participants that reported living with a *dog* evidenced a higher level of *Happiness* relative to those living with a *cat*. Lastly, in terms of magnitude the partial Eta squared effect size estimate indicated that the independent variable *Do You Live with a Dog or Cat?* evidenced a medium to large .09 (Eta squared effect size small, medium, large, = 0.01, 0.06, 0.14, respectively) effect on the dependent variable *Happiness*. Within the context of the full multivariate linear regression model, the strongest predictor of the dependent variable *Happiness* was the covariate predictor variable *Income Level*. This was evident as the covariate predictor *Income Level* evidenced the strongest standardized beta within the model (β=.39), as well as the largest effect size (PES=.65).

Figure 4.119   What multivariate analysis told us for sample study one

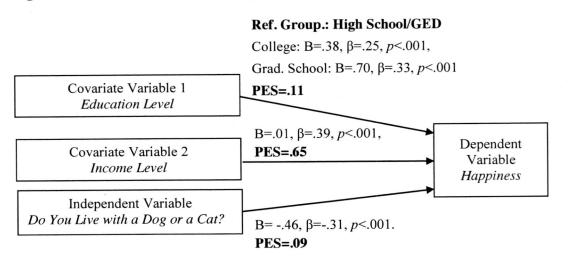

4.8 STEP 7: WRITE-UP & REPORT

The write-up and report is a process that might seem as enigmatic as the process of data analysis. There are many sources to refer to professional style (APA, AMA, MLA) to which an article must conform. However, there do not seem to be many sources that inform the aspiring author of what topic to discuss first in an article, to discuss second, and so on. Therefore, while I will not present a full lecture, I would like to share a few techniques I came up with for publishing research reports, particularly peer-reviewed articles. Please keep in mind that I have a rather minimalistic data analytic perspective (as opposed to heavy theory), where I aim to effectively present study constructs, what past research suggests regarding those constructs, and the potential contribution represented by the original study presented in the paper.

4.8.1 Body of the Paper

You will see below, we have drafted a light paper outline following the APA style guidelines of writing. Please note that the instructions are rather short to facilitate an easier application.

4.8.1.1 Abstract

We begin with an abstract page that include the page header, as well as the word "Abstract" centered atop with no bold, formatting, italics, underlining, or quotation marks. The abstract (non-indented, double spaced) should be between 150-250 words. Here we will present a concise summary of the key points of your research. Specifically, although there is not a need to make definitive statements ("The research question is…"), we would at least refer to the research topic, research questions, participants, methods, study design, results, data analysis, and conclusions, as well as possible implications for future research. The focus of and the information mentioned will vary by the specific study (e.g., some studies do not have a research question to mention). You may also want to specify *Keywords:* at the bottom.

4.8.1.2 Introduction

The introduction should present a brief but compelling statement regarding the study topic and a clear rationale for why the issue needs to be researched. A strong introduction may describe the prior theoretical and empirical research, as well as the current knowledge regarding the study topic with clear citations. For example, you can see in out paper below, essentials, we first substantiate the significance of the construct Happiness "The study of *Happiness* is a well-researched area that has been related to several outcomes, including education level (citation), income level (citation) and several other factors (citation)." Then we identify a need for research "However, level of Happiness has not been examined in relation to living with a dog or cat." Next, we describe how our study attempts to meet this research need "Therefore, the current study will examine if Happiness scores differ at a significant level, among study participants that live with a dog or cat."

4.8.1.3 Literature Review

The literature review needs to be focused and finite. Many study variables have an extensive literature, so if you are not focused there are many directions that could be taken. Therefore, the literature review must be approached with a clear agenda that protects from veering in too many directions. In our sample paper (refer to **Section 4.8.2 Sample Manuscript for Sample Study One**), you will see a simple four point formula for writing a literature review that is a bit of simplification that won't meet the needs of every study, but it is a decent place to start. This is the format I use to begin a literature review, and then adjust as needed.

The section describes that the first step is to discuss the dependent variable (**1st** *Discuss the Dependent Variable Construct*). For example, below we describe several general and specific characteristics of the dependent variable *Happiness*. Specifically, we discuss the construct *Happiness* by definition, as well as measurement with clear citations. We then discuss outcomes related to the dependent variable (absent of any discussion regarding the independent variable). This illustrates many things, including the far reaching impact of the dependent variable.

Next, we discuss the independent variable (**2nd** *Discuss the Independent Variable Construct*). Here we might describe the same elements we mentioned regarding the dependent variable (e.g., definitions, measurement, outcomes linked to the variable). We might also mention that the independent variable may not have been examined in a certain context that sets up the connection between the independent and dependent variables.

For example, in the sample manuscript we state "Although pet ownership has been linked to several outcomes, research has not been conducted that examines how these outcomes differ based upon pet type. Specifically, data have not been gathered that reflects if outcomes might differ if individuals reside with a dog, cat,..." Notice, although we didn't mention the dependent variable *Happiness* definitively, we inferred there is a need to examine the independent variable in relation to "outcomes," which our dependent variable certainly is.

Next, we have a paragraph that discusses the relationship between the independent and dependent variables (**3rd** *Discuss the Link between the Dependent and Independent Variables*). Here we would discuss the research, and related research, that has been conducted that relates these two constructs. However, it is essential to also describe the research that has not been conducted examining these two variables, which might be the focus of your study. For example, you will notice the first sentence states "Research has not been conducted that examines if the emotion happiness occurs on a more or less frequent basis relative to the type of pet an individual selects among the general population."

The final paragraph before the Methods section has a precise function. Specifically, in this paragraph we will enter the information from our study map (**4th** *In the Final Paragraph Before the Methods, Enter the Study Map Information*). Recall, the study map lists the research question and hypothesis, as well as the covariate, independent, and dependent variables. You will notice in this paragraph below, here we describe the independent and dependent variable relationship in our study question "The current study will examine if people who live with dogs have a significantly higher level of happiness relative to people who live with cats."

We also mention the covariate variables we will include in analysis "The analysis will also include several covariate variables that have been shown to impact happiness, such as education and income levels." We could have also added our hypothesis if desired. Lastly, it is a good idea to end on a compelling note where the final sentence before the methods reminds the reader why this analysis is important "This analysis represents the first attempt at gathering preliminary data that might suggest the impact of pet ownership might vary by pet type."

4.8.1.4 Methods

Ideally, the Methods section would be written with a level of clarity that would allow another researcher to duplicate the study. This information is essential as it is the basis for the reader to observe how the study sample might reflect the target population to judge the degree to which study results may be generalizable. Specifically, the Methods section should present the *following items:*

Recruitment of Study Participants

The Methods should describe where the study participants came from. Also, any attrition (i.e., study participants who exit the study prior to completion) of study participants should be reported. Below, we describe that the sample study drew the sample from a single apartment complex.

Informed Consent and IRB Approval

The Methods should describe how the process of informed consent was implemented, as well as how IRB approval was secured. Both these elements are described in the sample manuscript following this section.

Data Collection

The precise manner in which data were collected should also be described.

Noting whether conditions were manipulated or naturalistic

The individuals in our sample study are drawn from a group of community residents in a naturalistic setting. However, any type of manipulated conditions (e.g., an intervention that uses random assignment of cases to each study condition) should be described.

A description of the sample (e.g., demographic characteristics)

The sample description presents important characteristics of the study participants, such as demographic characteristics (i.e., race, gender, age, education). In our study, we present the demographic variables that were collected (i.e., *Education Level*, *Income Level*).

Group assignment

In our sample study, we did not assign study participants to groups, as might be done in a randomized control trial per se. However, any group assignment, including a description of a random or non-random design, should be mentioned and described.

Measurement instruments

The description of measurement instruments should include references, as well as available reports of the reliability and validity evidenced by the measurement tool in prior research. At times there may not be any materials along this line to report. However, at minimum, as we did regarding our measure of *Happiness* described below, for a continuous measure report the number of items in the measure (*e.g., our Happiness measure is a 5-item scale*) and the continuum that items are measured along (*e.g., the items within our Happiness measure are measured along a 5-point continuum where 1= Strongly Disagree and 5= Strongly Agree*).

The method used to compute the scale (e.g., taking the mean score) should be described as well as the potential range of scale scores (e.g., 1-5), and what higher scores indicate (e.g., higher scores = Higher *Happiness*). Lastly, the internal consistency reliability (e.g., Cronbach's Alpha) should be reported. For example, in our sample manuscript we report the Cronbach's alpha for the 5-item scale measuring *Happiness* in the current sample is .89.

Research design

The research design used in the study should always be described. For example, in our sample manuscript we mention this study incorporates a cross-sectional study design.

Data analysis plan

Data analysis is a process. The data analysis plan describes how the process is approached in each data analysis study. For example, in the sample manuscript included at the end of this section we describe that in the current data analysis plan, analysis will be conducted in three phases. Then the three phases are presented, which are essential **Step 4: Univariate Analysis**, **Step 5: Bivariate Analysis**, and **Step 6:**

Multivariate Analysis. Furthermore, in the second part of this section we describe the process and results of **Step 3: Checks of Data Integrity**.

4.8.1.5 Results

You will see in the sample manuscript, the results are presented in a systematic manner that follows the data analysis plan. Please note that the first section is Descriptive Analysis, where we describe the study variables via univariate statistics. The next two sections present the results of statistical testing in the Bivariate Analysis and Multivariate Analysis sections. In addition to text, the findings are also presented in tables that present a summary of the data in a more user-friendly format. The appropriate (APA, AMA, MLA) *Publication Manual* should be observed for the proper method of reporting.

4.8.1.6 Discussion

In the Discussion section, the author should evaluate and interpret the findings of the study. The Discussion sections should begin with a statement of support or nonsupport for the study hypotheses (if applicable) relative to the study findings, as we did in our sample manuscript. Typically, if a hypothesis is not supported, an author will describe possible explanations of the findings. Explanations often consider that a non-supported hypothesis might be a true finding, as well as the possibility that the result was a product of some weakness in the study design (e.g., imprecision of measures, small study sample size), or an alternate hypothesis. This section also incorporates a limitations section where shortcomings in the study design or some other feature of the study might be discussed.

4.8.1.7 Conclusion

As you can imagine, the conclusion should be a reasonable and accurate statement based upon implications of the findings. It should be a statement of what the findings infer, but should be careful not to go beyond those boundaries. For example, the conclusion within our sample manuscript based upon the study findings is that pet owners that live with dogs may have a higher degree of *Happiness* relative to those who live with cats. The conclusion also suggests that future research should be conducted that asks study participants probing qualitative questions that might better identify the reasons behind this finding. The conclusion does not go beyond the boundaries of the study implications to recommend that everyone who owns a cat should trade their pet in for a dog if they want to be happier.

4.8.2 Sample Manuscript for Sample Study One

THE RELATIONSHIP BETWEEN HAPPINESS AND LIVING WITH A DOG OR CAT

Abstract

The current study examines if participants living with a dog display significantly higher levels of happiness relative to those living with a cat. The study employs a cross-sectional design and involves a convenience sample of residents living in an apartment complex in a major northeastern city. A multivariate linear regression model indicated that happiness scores were significantly associated with education and income level. Furthermore, the regression model reflected that study participants that live with a dog have significantly higher levels of happiness relative to those that live with a cat. Future research would benefit from a mixed-methods design that might include a qualitative piece capturing participant perspectives regarding why dog ownership might be associated with higher happiness scores relative to cat ownership.

Keywords: Happiness, Pets, Dogs, Cats

INTRODUCTION

(An Overview of the Central Issue and Reason for the Study)

The study of happiness is a well-researched area that has been related to several outcomes, including education level (*citation*), income level (*citation*) and several other factors (*citation*). However, level of happiness has not been examined in relation to living with a dog or cat. Therefore, the current study will examine if happiness scores differ at a significant level, among study participants that live with a dog or cat.

LITERATURE REVIEW

(1st *Discuss the Dependent Variable Construct)*

Happiness is often measured through participant reports of that specific emotion (*citation*). However, happiness has also been studied in relation to general life satisfaction (*citation*) and elevated mood (*citation*)....

(Discuss Outcomes Linked to the Dependent Variable,

but not the Independent Variable Yet)

One of the most consistent predictors of happiness is income level (*citation*). Specifically, numerous studies have linked higher levels of income to significantly higher levels of happiness (*citation*)...

Happiness has also been linked to higher educational attainment (*citation*). Smith and colleagues (*citation*) identified that....

(**2ⁿᵈ** *Discuss the Independent Variable Construct*)

Pet ownership has been linked to several outcomes (*citation*). For example, elderly residents of a specialized living facility reported higher levels of general life satisfaction when a pet resided with them in their apartment (*citation*). Similar findings are reflected with other age groups.....

Although pet ownership has been linked to several outcomes, research has not been conducted that examines how these outcomes differ based upon pet type. Specifically, data have not been gathered that reflects if outcomes might differ if individuals reside with a dog or cat,...

(**3ʳᵈ** *Discuss the Link between the Dependent and Independent Variables*)

Research has not been conducted that examines if the emotion happiness occurs on a more or less frequent basis relative to the type of pet an individual selects among the general population. However, there is evidence to suggest such a relationship might exist where a group of pet owners with pets that demanded more care, such as cats, evidenced higher stress levels relative to those who owned more independent animals, such as dogs (citation). Specifically,...

(**4ᵗʰ** *In the Final Paragraph Before the Methods, Enter the Study Map Information*)

The current study will examine if people who live with dogs have a significantly higher level of happiness relative to people who live with cats. The analysis will also include several covariate variables that have been shown to impact happiness, such as education and income levels. This analysis represents the first attempt at gathering preliminary data that might suggest the impact of pet ownership might vary by pet type.

METHODS

Study Participants Recruitment

Study participants were recruited in a single apartment complex in a large northeastern city. Specifically, during a tenant meeting of approximately 200 residents were invited to participate in the current research. Subsequently, 100 residents of the apartment complex attending the tenant meeting accepted this invitation and completed the study survey. All 100 residents that composed this community sample were advised of and signed forms relating to informed consent. Furthermore, this project was approved by the IRB board at City Hospital.

Research design

The current study involved a cross-sectional design used to gather data from a non-probability non-random convenience sample of community residents.

Sample

The current research involved a sample of 100 community residents. Community residents were rather evenly dispersed regarding their education level, as 32% (n=32) indicated a highest level of education of high school/GED, while 29% (n=29) indicated the college level and 39% (n=39) indicated a graduate school level of education. The average study participant was approximately 21 years of age (M=21.72; SD= 2.55; MIN/MAX=18.00/26.00).

Measurement instruments

Demographics. A short scale examining study demographics (e.g., education and income levels) was created for the current study.

Alcohol use. Alcohol use is measured with the single item Have you ever had a drink of alcohol? (Yes/No). Study participants that indicate a response of Yes are also asked to indicate the number of drinks consumed in the last week.

Pet ownership. Pet ownership is examined via a single item where study participants are asked to indicate if they live with either a dog or cat.

Hobbies. Study participants are asked to describe their hobbies by indicating if they enjoy, skiing, surfing, and/or reading (circle all that apply).

Happiness. The construct happiness is measured using a 5-item scale created for the current study. Items are measured along a Likert-type scale using a continuum of 1 (Strongly Disagree) to 5 (Strongly Agree). The scale is computed through taking the mean score of valid responses, resulting in a potential range of scores of 1-5, with higher scores indicating a greater degree of happiness. A reliability analysis indicated the level of scale internal consistency was acceptable for the current sample (Cronbach's alpha=.89).

Data analysis plan

Data analysis will be conducted in three phases. First, all data will be analyzed descriptively via univariate analysis. Second, the relationship between the dependent variable happiness with the independent (do you live with a dog or cat?) and covariate (education level, income level) variables will be examined using bivariate analysis. Predictor variables (independent/covariate variables) associated with the dependent variable at a statistically significant level ($p<.05$) will be entered in the final multivariate model. Third, a multiple linear regression model will be used to model scores reflecting the dependent variable happiness as a function of all predictor variables included in the multivariate model.

Prior to data analysis an examination of test assumptions indicated a satisfactory level of homoscedasticity, linearity, and normality. Furthermore, data indicated that multicollinearity did not present a significant problem. A power analysis indicated 85 study participants would provide sufficient statistical power to detect a medium size effect between the independent and dependent variables within a multiple linear regression model using four predictors. Thus, the current sample of 100 study participants should provide sufficient statistical power.

Data were initially collected from 105 study participants. Of these, 10 evidenced missing data values, all of which were among the 5-item scale measuring happiness. Five of these study participants were excluded from analysis because less than 80% of items contained valid responses, which left a final sample of 100 study participants for analysis. The other five study participants provided valid responses for 80% of scale items (4 of the 5 scale items) and were included in the analysis using the mean score of the valid responses. Bivariate analysis indicated data appeared to be missing at random, as study participants with and without missing data values did not differ significantly by key characteristics, including education level, $X(2)=.41$, $p=.82$, and income level, $t(103)=.36$, $p=72$.

RESULTS

Descriptive Analysis

Table 1 presents a descriptive analysis of the variable do you live with a dog or cat? Data indicated that approximately half of study participants reported living with a dog ($n=47$; 47%) or cat ($n=53$; 53%). Table 2 presents a descriptive analysis of the variable happiness. Data indicated that the score reflecting happiness is 2.73 ($SD=.76$; MIN/MAX=1.20-4.40).

Bivariate Analysis

Table 3 presents a bivariate analysis of happiness scores by education level and do you live with dog or cat? A One-Way ANOVA indicated a statistically significant difference between mean scores reflecting happiness, $F(2, 97)=25.96$, $p<.001$. A Bonferroni post hoc test indicated the mean score for happiness differed where: 1) The score for high school/GED, $M=2.17$, ($SD=.70$) was significantly lower than college and graduate school scores; 2) The score for college, $M=2.70$ ($SD=.61$) was significantly higher than high school/GED, but lower than graduate school scores; and 3) The score for graduate school, $M=3.23$ ($SD=.55$) was significantly higher than high school/GED and college scores.

An independent-samples t-test reflected that the mean score reflecting happiness was significantly higher among study participants that reported living with a dog

(M=3.02, SD=.79) relative to those that reported living with a cat (M=2.48, SD=.62), $t(98)$=3.76, p<.001.

Table 4 presents a bivariate analysis of happiness scores with income level. Specifically, zero-order Pearson's correlation indicated that income level and happiness scores were positively correlated at a statistically significant level, $r(98)$=.88, p<.01.

Multivariate Analysis

Table 5 presents a multiple linear regression analysis examining scores reflecting happiness. Data indicated that the overall model was statistically significant, $F(4, 95)$=107.92, p<.001. Furthermore, the model explained over 80% (R =.82, Adjusted R =.81) of the variance in the dependent variable happiness.

In terms of individual predictors, data indicated that study participants with a college level of education (B=.26, SE=.09, β=.16, p<.01) had significantly higher happiness scores, in reference to those with a high school/GED level of education. Furthermore, data indicated that study participants with a graduate school level of education (B=.33, SE=.10, β=.22, p<.001) had significantly higher happiness scores, in reference to those with a high school/GED level of education. Higher scores reflecting happiness were also associated with higher levels of income (B=5.44, SE=.00, β=.74, p<.001), as well as living with a dog (B=-.23, SE=.07, β=-.15, p<.01) relative to a cat.

DISCUSSION

The study findings supported the study hypotheses that postulated that study participants that live with dogs have significantly higher levels of happiness relative to those that live with cats. There are several possibilities for this finding...

Limitations

This current study also incorporates several limitations. For example, the study design did not allow for study participants that lived with both a dog and cat...

CONCLUSION

The finding that pet owners that live with a dog evidenced a significantly higher degree of happiness relative to those that live with cats, suggests that future research should examine this topic further. Specifically, future research might benefit from incorporating a qualitative piece that includes probing qualitative questions that might better identify the reasons behind this finding...

Table 1

Descriptive Analysis of the Variables Education Level and Do You Live with Dog or Cat (*N*=100)

Variable	n	%
Education Level		
High School/GED	32	32.0
College	29	29.0
Graduate School	39	39.0
Do You Live with a Dog or Cat?		
Dog	47	47.0
Cat	53	53.0

Table 2

Descriptive Analysis of the Variables Income Level and Happiness (*N*=100)

Variable	M (SD)	MIN/MAX	Potential Scores
Income Level	197, 250.00 (103,336.79)	100,000.00-495,000.00	NA
Happiness	2.73 (.76)	1.20-4.40	1.00-5.00

Table 3

Bivariate Analysis of Happiness Scores by Education Level and Do You Live with Dog or Cat (N=100)

Variable	M (SD)	t/F (df)	p
Education Level*		25.96 (2, 97)	.000
High School/GED	2.17 (.70)		
College	2.70 (.61)		
Graduate School	3.23 (.55)		
Do You Live with a Dog or Cat?		3.76 (98)	.000
Dog	3.02 (.79)		
Cat	2.48 (.62)		

*Bonferroni post hoc test indicated the mean score for Happiness differed where: 1) The score for High School/GED was significantly lower than College and Graduate School scores; 2) The score for College was significantly higher than High School/GED, but lower than Graduate School scores; and 3) The score for Graduate School was significantly higher than High School/GED and College scores.

Table 4

Bivariate Analysis of Happiness Scores with Income Level ($N=100$)

Variable	1	2
1. Income Level	--	.88**
2. Happiness		--

**p<.01*

Table 5

Multiple Linear Regression Analysis Examining Happiness Scores ($N=100$)

Variable	B	SE	β	p
Education Level				
High School/GED (Reference Group)				
College	.26	.09	.16	.003
Graduate School	.33	.10	.22	.001
Income Level	5.44	.00	.74	.000
Do You Live with a Dog or Cat?	-.23	.07	-.15	.002

Note. For Model: $R = .82$, Adjusted $R = .81$, $F(4, 95) = 107.92$, $p < .001$.

PART 5

A QUANTITATIVE STUDY WITH A CATEGORICAL DEPENDENT VARIABLE

5.1 What Will This Section Tell Us?

In Part 5, through sample study two, we will review how to conduct a quantitative study that examines a categorical, specifically a dichotomous, dependent variable. As mentioned earlier in this text, although there are seven steps involved in a data analysis study, the procedures and details within each step vary according to the level of measurement (e.g., categorical, continuous) of the dependent variable. Therefore, Part 5 will not only illustrate how to apply *The 7 Steps of Data Analysis* model toward completing a study with a dichotomous dependent variable, but also how the analysis differs relative to sample study one where a continuous dependent variable was examined. For example, similar to our first sample study, we will apply *The 7 Steps of Data Analysis* toward completing a data analysis study examining if people living with a dog or a cat have significantly different levels of *Happiness*. However, where in sample study one, *Happiness* was measured on a continuous scale (1-5), in sample study two the dependent variable will be measured as a dichotomous variable where study participants are coded as *Happy* (Yes or No).

In a way, sample study two will be an extension of sample study one as the same variables are used, with the exception of the dependent variable being dichotomized. The practice of dichotomizing a continuous dependent variable and repeating an analysis with the same predictor variables is not uncommon in the field of quantitative research. For example, many continuous scales have a meaningful cutoff score. Subsequently, it is often appropriate to analyze the outcome as a continuous variable or a variable dichotomized using the cutoff score, as we will do in sample study two regarding the dependent variable *Happiness*.

The term *cutoff score* (sometimes called a boundary score) refers to a score that is selected to represent the boundary between a typical and a significant level of an outcome surveyed via a study measurement tool. Cutoff scores are common in mental health research toward identifying clinically significant levels of depression (e.g., significantly depressed: Yes/No), anxiety, behavioral problems, and so on. Regarding our study measurement tool for the dependent variable *Happiness*, we will use a cutoff score to identify values that reflect a significant degree of *Happiness* (*Happy*=Yes) relative to all other scores (*Happy*=No).

Many times, a study measurement tool has a normed cutoff score, which typically means the creators of the scale applied a method to determine the cutoff score value. For example, one method of calculating a cutoff score is to consider the mean score and standard deviation value among both a clinical (e.g., patients being treated for clinically significant levels of *depression*) and non-clinical sample (e.g., a community sample of non-depressed individuals), then estimate the score at which a study participant has a greater probability of belonging to the clinical rather than the non-clinical sample (Jacobson & Truax, 1984). Subsequently, in this case you would not need to identify the cutoff score, as the value would be specified by the creators of the instrument.

However, our composite measure of *Happiness* does not have a cutoff score, so we will need to select a value. A common method would be to use the mean and standard deviation values of the continuous dependent variable *Happiness* (see **SPSS Output 4.25**) to identify high levels of this construct. For example, we could define high levels of *Happiness* as scores one half a standard deviation (SD=.76 divided by 2=.38) above the mean *Happiness* score. Recall, the mean score reflecting *Happiness* is 2.73, so scores half a standard deviation above would be 3.11 (a mean of 2.73 + ½ SD of .38 = 3.11). Thus, *Happiness* scores above the value 3.11 will be coded as *Happy* (Yes) and values of 3.11 or less will be coded as *Happy* (No).

After dichotomizing the dependent variable, we will apply *The 7 Steps of Data Analysis* model to reveal the factors that explain a greater likelihood of a study participant being *Happy* (Yes) or not *Happy* (No). It would be redundant to repeat every procedure completed in the first sample study here, such as entering, cleaning, and coding the data. Therefore, this section (Part 5) will be somewhat truncated. However, for readers who are reading sample study two before one, we will note when and where to refer to important materials already presented in sample study one.

5.2   Step 1: Study Map

Step 1 (Study Map) is the stage of the quantitative analysis where the study variables and hypothesized variable relationship(s) are listed in both text (**Section 5.2.1**) and diagram (**Section 5.2.2**) form.

5.2.1   The Study Map in Text

We will now present our Study Map for sample study two, which will specify the study research question, hypothesis, dependent variable, independent variable, and covariate variables. Specifically, in sample study two these elements are:

Research Question: Are People Who Live with Dogs More Likely to be Happy Relative to People Who Live with Cats?

Hypothesis: People Who Live with Dogs Are More Likely to be Happy Relative to People Who Live with Cats.

Dependent Variable: Happy (Yes or No)

Independent Variable: Do You Live with a Dog or a Cat?

Covariate Variables: Education Level, Income Level

5.2.2   The Study Map in a Diagram

We can also present the Study Map in diagram form, as presented in **Figure 5.1**.

Figure 5.1   The variable relationships under examination in sample study two

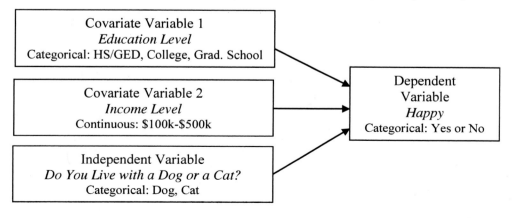

5.3   Step 2: Data Entry

Sample study two requires only one added task regarding **Step 2: Data Entry**, relative to the data coding, entry, cleaning, and recoding completed in sample study one. Specifically, sample study two requires an additional variable recode where *Happiness* scores are changed from a continuous format, into a categorical (dichotomous) format, which will be presented in **Section 5.3.1**.

5.3.1   Recoding: Making Happiness Happy

Recall, in our earlier computation we identified that *Happiness* scores above the value 3.11 will be recoded as *Happy* (Yes) and values of 3.11 or less will be recoded as not *Happy* (No). To perform this recode in SPSS go to:

1) **Transform→ Recode Into Different Variables→**.

2) Use the center arrow to move the variable to be recoded, *Happiness*, from the left hand column into the right hand column.

3) In the box to the right labeled **Name**, enter the name of the new transformed variable, which for sample study two is *Happy*.

4) Click **Change**, which will move the new variable name into the center column, as presented in **Figure 5.2**.

Figure 5.2   Recoding the dependent variable *Happiness* into *Happy* (Yes or No)

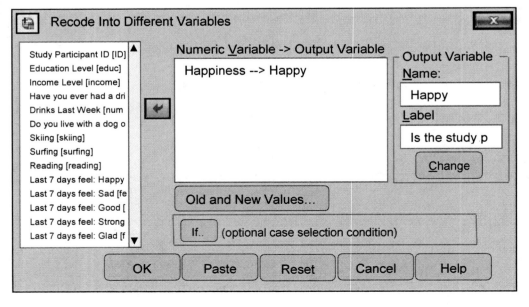

5) Click **Old and New Values**.

6) Recall, the full range of scores for *Happiness* is 1.00-5.00 and the cutoff score for *Happy* (Yes) is above 3.11. To dichotomize the scale, we must code from the minimum score to the maximum score of the *Happiness* scale, while incorporating the cutoff score of 3.11. Therefore, scores reflecting *Happy* (Yes) will be 3.12-5.00, while scores reflecting *Happy* (No) will be 1.00-3.11.

7) In the dialogue box that opens, within the left hand column labeled **Old Value**, click **Range**.

8) Enter the range of values reflecting *Happy* (No), which is 1.00 to 3.11.

9) Within the right hand column labeled **New Value**, click **Value**, and enter a '0' (0=No) in the cell, click **Add**.

10) Within the left hand column labeled **Old Value**, click **Range**.

11) Enter the range of values reflecting *Happy* (Yes), which is 3.12 to 5.00.

12) Within the right hand column labeled **New Value**, click **Value** and enter a '1' (1=Yes) in the cell, click **Add** (as presented in **Figure 5.3**).

Figure 5.3 Recoding the dependent variable *Happiness* into *Happy* (Yes or No)

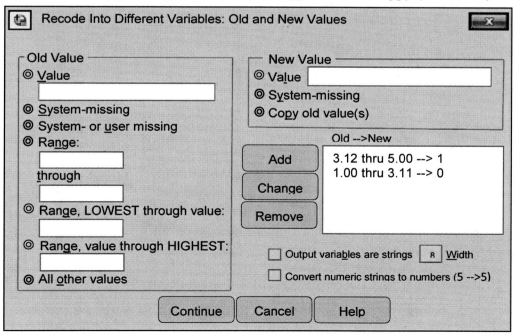

13) Click **Continue**.

14) Click **OK** or **Paste**.

15) Next, go into the **Variable View** portion of the SPSS database where you should find the newly created variable at the end of the list of variables.

16) Under **Label**, enter a description for the new dependent variable, such as *Is the study participant happy?*

17) Under **Values**, enter a '1' for Yes and a '0' for No.

18) Check to be sure the recode was successful through conducting the frequencies procedure to examine the new dependent variable *Happy*. Recall, to conduct the frequencies procedure in SPSS go to:

    a. **Analyze→ Descriptive Statistics→ Frequencies**.

    b. In the dialogue box that opens (see **Figure 4.25**), move the variable *Happy* from the left hand column into the right hand column.

    c. Click **OK** or **Paste**.

SPSS Output 5.1 presents the results of the frequencies procedure for the recoded dependent variable *Happy*. The output describes there are 69 (69%) scores below the cutoff point of 3.11 or less, as well as 31 (31%) cases above the cutoff point.

SPSS Output 5.1

Is the study participant happy?

		N	Percent	Valid Percent	Cumulative Percent
Valid	No	69	69.0	69.0	69.0
	Yes	31	31.0	31.0	100.0
	Total	100	100.0	100.0	

To check the accuracy of this recode, we can observe the frequency distribution of *Happiness* scores in their continuous form. To do this, conduct the frequency procedure using the variable *Happiness* in SPSS through going to:

    a. **Analyze→ Descriptive Statistics→ Frequencies**.

    b. In the dialogue box that opens (see **Figure 4.25**), move the variable *Happiness* from the left hand column into the right hand column.

    c. Click **OK** or **Paste**.

SPSS Output 5.2 presents the frequency table, which will be included in the results of the frequencies procedure. We can check our recode by observing if within the original distribution, there is a 31/69 study participant split above and below the cutoff score of 3.11. Within the output, the location of the cutoff score of 3.11 is indicated using a dotted line. We can see there are 69 (69%) scores above the dotted line, which reflect the scores categorized as not *Happy* (Happy=No; between 1.00 and 3.11), as well as 31 (31%) scores below the dotted line, which reflect the scores categorized as *Happy* (Happy=Yes; between 3.12 and 5.00). Thus, as the recoded dichotomous dependent variable *Happy* also identifies a group of 69 and 31, the recode appears successful.

SPSS Output 5.2

Happiness (Continuous Scores)

		Frequency	Percent	Valid Percent	Cumulative Percent
	1.20	2	2.0	2.0	2.0
	1.40	3	3.0	3.0	5.0
	1.60	4	4.0	4.0	9.0
	1.80	2	2.0	2.0	11.0
	2.00	9	9.0	9.0	20.0
Happy	2.20	10	10.0	10.0	30.0
(No)	2.40	12	12.0	12.0	42.0
	2.60	12	12.0	12.0	54.0
↑	2.80	5	5.0	5.0	59.0
3.11	3.00	10	10.0	10.0	69.0
Cutoff	3.20	5	5.0	5.0	74.0
Score	3.40	10	10.0	10.0	84.0
↓	3.60	6	6.0	6.0	90.0
Happy	3.80	1	1.0	1.0	91.0
(Yes)	4.00	4	4.0	4.0	95.0
	4.20	3	3.0	3.0	98.0
	4.40	2	2.0	2.0	100.0
	Total	100	100.0	100.0	

But don't stop there! Checking scores in the raw data is always a good idea. In order to make sure our recode was successful in this manner, begin by:

1) Checking the raw data, by lining up the original (*Happiness*) and recoded (*Happy*) variables next to each other in the **Data View** portion of the SPSS database (to move a variable, right click on the variable, then right click again and hold down the button, then drag the variable across the other variables).

2) Then right click on the top of the original variable (*Happiness*) and in the menu that opens click **Sort Descending**. Then scroll down to find the value that divides the cutoff score (3.11) in the original continuous variable *Happiness*. Next, observe if the two categories (Yes and No) within the recoded variable (*Happy*) switches from Yes to No at the cutoff point of 3.11 within the original variable *Happiness*.

3) For example, **Figure 5.4** presents the **Data View** portion of the SPSS database where the dependent variable *Happiness* scores cross the cutoff point of 3.11, as well as the point where the coding for the new dependent variable *Happy* switches from Yes to No.

4) We see the categorical dependent variable *Happy* switches from Yes to No at the point where the *Happiness* scores cross the value 3.11. You may notice there is no actual *Happiness* score of 3.11, but instead an interval of scores enveloping the score of 3.11, which is 3.00 and 3.20. Since there is no actual *Happiness* score of 3.11, we will observe if the Yes and No coding for the variable *Happy* switches at this interval, where the cutoff point 3.11 would be.

5) Via comparing the two variables, *Happiness* and *Happy* in **Figure 5.4**, we can see that the Yes and No coding for the variable *Happy*, does switch at the cutoff score of 3.11 regarding the variable *Happiness*, suggesting a successful recode.

Figure 5.4   Checking the raw data for between *Happiness* and *Happy*

	Happiness ↓	Happy ↓	
30	3.20	1.00	
31	3.20	1.00	1 = Happy (Yes)
32	3.00	0.00	0 = Happy (No)
33	3.00	0.00	

Happiness: > 3.12

Happiness: ≤ 3.11

5.4   Step 3: Checks Of Data Integrity

Just like in sample study one, **Step 3: Checks of Data Integrity** is concerned with one central question: **Are the Data Appropriate for Data Analysis?** It is also worth repeating that this is perhaps the most important question of any study, even more so than the research question(s) and hypothesis(es). Why? Essentially because if the data are not appropriate for data analysis, the research question(s) and hypothesis(es) are likely to be answered incorrectly!

As described in **Section 4.4** within sample study one, there are four main areas in which data integrity must be checked. In this section, you will see that there is variation in the procedures required in the first two checks of data integrity (specifically the test of statistical power and test assumptions) between the first and second studies. However, the procedures employed for the second two checks of data integrity (Missing Data and Study Measurement Tools) remain the same. Subsequently, let's discuss our four checks of data integrity as each pertains to our second sample study, including checks of statistical power (**Section 5.4.1**), test assumptions (**Section 5.4.2**), missing data (**Section 5.4.3**), and measurement tools (**Section 5.4.4**).

5.4.1   Statistical power

Please see the corresponding section **4.4.1 Statistical power** within sample study one for a full description of statistical power. Section 4.4.1 also details how the power analysis is often the primary tool in statistics used to estimate the sample size needed to generate enough statistical power to examine the relationships between study variables. In this section, rather than discuss these materials a second time, we will apply this information to conduct a power analysis for sample study two.

5.4.1.1 Power Analysis

Please see the corresponding section **4.4.1.1 Power Analysis** within sample study one for a full description of the power analysis. Recall, in this section, we described when performing a power analysis we must provide the following settings:

Power: We must set the value of power, which is defined as the ability to detect a significant effect if a real difference exists. For sample study two, we will set power at the standard accepted value of .80 (80%).

Alpha: Alpha refers to the level of probability we are willing to accept that the findings produced are by chance. For sample study two, we will set alpha at the standard acceptable value of .05.

Effect Size: The 'effect size' refers to the magnitude of the effect between variables. Effect sizes are categorized as being small, medium, or large. For sample study two, we will specify a medium 'effect size.'

Sample Size: Once the power, alpha, and the effect size are set, we will use the statistical software to calculate the required sample size to generate enough statistical power to test the relationship between variables.

Power Analysis Using the Power and Precision Software Program. For this power analysis, we will use the statistical software program *Power and Precision*. This is a specialized software program designed solely for conducting various types of power analysis procedures. Once the Power and Precision software program is open, there will be an option in the upper left hand corner to open a new project. Once a new project is selected the dialogue box presented in **Figure 5.5** will open. You will see within the dialogue box there are various options available that can be selected based upon the inferential statistical procedure (e.g., chi-square, binary logistic regression) for which a power analysis is needed.

For example, we need to perform a power analysis for the binary logistic regression model that will be used in multivariate analysis (**Step 6**) in sample study two. Our independent variable (predictor) *Do You Live with a Dog or Cat?* is a categorical (dichotomous) predictor with a two category response (Dog or Cat). Therefore, as presented in **Figure 5.5**, we would select the toggle box next to the term **Logistic regression, one categorical predictor (two levels)**. After the appropriate category is selected within this dialogue box, click the button marked **Next** at the bottom of the page (not shown in **Figure 5.5**).

Figure 5.5   Power analysis for logistic regression with one categorical predictor

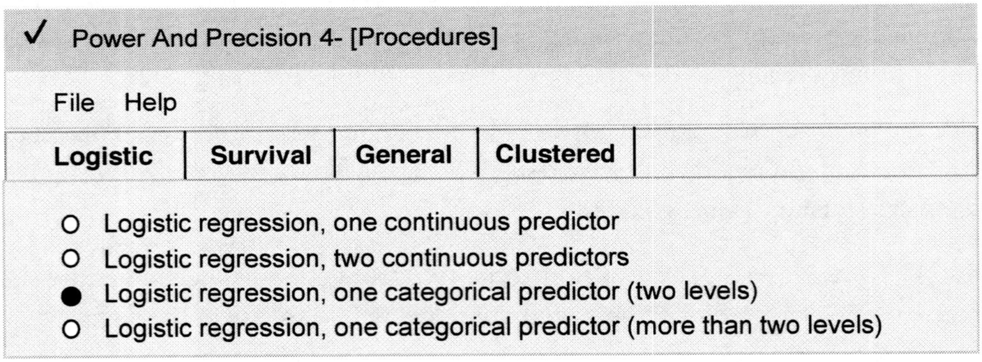

Subsequently, the dialogue box presented in **Figure 5.6** will open. Within this figure:

1) Refer to the column to the left with the title **Predictor Variable**. Under the term **Group Name** there are two cells where a name for each level of the predictor variable (here our independent variable) can be entered. Here we have entered *Lives with Dog* and *Lives with Cat* in the first and second cells, respectively.

2) The next column, labeled **Relative Proportion**, reflects the relative proportion of study participants within each group. This proportion may be known or unknown, depending on the specific study. However, for our study, we will project an approximately equal number of study participants. Specifically, we will assume that of the 1.0 study participants, 50 will report living with a dog and 50 will report living with a cat. Therefore, based on these equal proportions, you will see the value 1.0 has been entered for each group.

These two numbers would be modified if the groups were unequal. For example, if twice as many study participants were in the first group relative to the second group, a 2.0 would have been entered in the first group and a 1.0 in the second group. Next, under the term *Reference Group*, group one *Lives with Dog* has been selected. This group has been selected mainly because the study examines if study participants that live with a dog are more likely to be happy in reference to those that live with a cat.

3) **Alpha**: In the lower left hand corner, we see that *Alpha* is set at .05 (Tails=2).

4) **Power**: Furthermore in the lower right hand corner we see *Power* is .80.

5) **Effect Size**: Refer to the two cells beneath the term **Event Rate** under the column marked **Outcome** to the right. The effect size between the predictor (independent variable) and dependent variables will be reflected by the specified event rate, which is the proportion of study participants expected to experience the outcome within each group. The event/outcome in sample study two is being *Happy* (Yes or No).

In the top cell, the value **.54** has been entered, indicating we expect **54% of study participants that live with a dog to be Happy (Happy=Yes)**. In the bottom cell, the value **.26** has been entered, indicating we expect **26% of study participants that live with a cat to be Happy (Happy=Yes)**. These estimates are hypothetical, but in actual research these values would most likely be based upon numbers identified in prior research studies.

The odds ratio (computed on the next page) based upon the difference between these event rates is approximately 3.3 (i.e., people that live with dogs are 3.3 times more likely to be happy relative to people that live with cats), which is a medium size effect. The Odds Ratio for the reference group is shown as 1.0 (and beta is shown as 0.0), which is set as a standard for the reference group. The corresponding values for the other group are shown in comparison to the reference group. These values are used to quantify the magnitude of the effect size between the predictor (independent variable) and dependent variables.

For example, our event rate projections correspond to numbers in the computations below as approximately 54% of the group that lives with dogs (which is 27 of 50 study participants) and 26% of the group that lives with cats (which is 13 of 50 study participants) will experience the event of being *Happy* (Happy=Yes). The odds ratio is the odds of the event occurring in one group divided by the odds of the event occurring in the other group, which for these data are expressed as:

Odds in group 1 (Lives w/Dog) =

with event (Happy=Yes) is **27**
without event (Happy=No) is **23**

$(27/23) = 1.174$

Odds in group 2 (Lives w/Cat) =

with event (Happy=Yes) is **13**
without event (Happy=No) is **37**

/ $(13/37) = 0.351$

$$1.174 / 0.351 = 3.34$$
Odds Ratio = 3.34

6) **Sample Size**: Lastly, we can see that with all other necessary considerations made in the power analysis, a sample size of 100 study participants is sufficient to provide enough statistical power to test the specified relationship between the independent variable and dependent variable.

Lastly, be sure to reference the piece **Why Only One Power Analysis?** In the corresponding power analysis section **4.4.1 Statistical power** in sample study 1.

Figure 5.6 Settings for the power analysis for logistic regression

	Predictor Variable		Outcome				
Group Name	Relative Proportion	Reference Group	Event Rate	Odds Ratio	Beta	Relative Risk
☑ Lives with Dog	1.0	●	0.54	1.00	0.00	1.00
☑ Lives with Cat	1.0	O	0.26	0.30	-1.21	0.48

Alpha=0.05, Tails=2   Total sample size 100   **Power** 80%

A sample size of 100 is required

5.4.2 Test Assumptions

As we said in sample study one regarding multiple linear regression, all quantitative models have certain assumptions that must be met before a statistical procedure can be applied. However, binary logistic regression has fewer test assumptions relative to linear regression. For example, binary logistic regression does not assume a linearity, a normal distribution of scores, or homoscedasticity; so there is not a need to test for these assumptions. However, there is still a need to test for multicollinearity, which will be discussed in this section.

5.4.2.1 Multicollinearity

Multicollinearity is a necessary check when conducting a binary logistic regression model with multiple predictors. The assessment of multicollinearity between the study predictor variables conducted in sample study one is also applicable to sample study two. Thus, there is not a need in sample study two to assess multicollinearity. Subsequently, please refer to the corresponding section **4.4.2.2 Multicollinearity** in sample study one for a detailed description of multicollinearity, as well as how to address challenges presented by this factor. There is one small *SPSS specific* tip regarding multicollinearity and logistic regression we should mention here. Recall, within the corresponding section **4.4.2.2 Multicollinearity** in sample study 1, we discussed assessing multicollinearity through the *collinearity diagnostics* function in the linear regression. It is important to note the SPSS software currently does not have a corresponding collinearity diagnostics function for binary logistic regression. However, if your study incorporates binary logistic regression, you can crossover to use the collinearity diagnostics function available in linear regression to estimate the multicollinearity between predictor variables.

5.4.3 Missing Data

A full discussion of the issues relating to missing data, as well as methods used to address missing values in a dataset, is presented in the corresponding section **4.4.3 Missing Data** in sample study one.

5.4.4 Measurement Tools

A full discussion of the issues relating to measurement tools, as well as the methods used to assess the properties of measurement tools is presented in the corresponding section **4.4.4 Measurement Tools** in sample study one.

5.5　Step 4: Univariate Analysis

Please refer to the corresponding section **4.5 Univariate Analysis** for a full discussion of this step. For the purposes of sample study two, the only univariate analysis needed is a frequencies procedure applied to the dependent variable *Happy* (Yes or No). Recall, we had conducted this frequencies procedure for the dependent variable *Happy* already in sample study two in section **5.2 Data Entry** while recoding the variable. This frequencies procedure indicated that of the 100 study participants, 69 (69%) are coded as not *Happy* (*Happy*=No) and 31 (31%) are coded as *Happy* (*Happy*=Yes).

5.6　Step 5: Bivariate Analysis

Please see the corresponding section **4.6 Bivariate analysis** in sample study one for a complete description of this step. As in sample study one, we will employ bivariate statistical procedures to identify the predictor variables associated with the dependent variable *Happy* (Yes or No) at a statistically significant level. The Bivariate Test Key (see **Figure 5.7**) instructs to use an independent-samples t-test to examine the relationship between the dependent variable *Happy* and covariate variable *Income Level*, as well as chi-square analysis to relate the other predictors (*Education Level* and *Do You Live with a Dog or a Cat?*) with the outcome variable.

Figure 5.7　The Bivariate Test Key (Study Variables Plugged In)

Variable 1	Variable 2	Test To Use
Categorical (2 categories)	Continuous	Independent Samples T-Test
Happy (Yes, No)	Income Level ($100k - $500k)	Independent Samples T-Test
Categorical (≥3 categories)	Continuous	One-Way ANOVA
Continuous	Continuous	Correlation
Categorical	Categorical	Chi-Square
Education (HS/G, C, GS)	Happy (Yes, No)	Chi-Square
Live w/Dog/Cat (Dog, Cat)	Happy (Yes, No)	Chi-Square

5.6.1　Independent Samples T-Test

The bivariate test key (**Figure 5.7**) indicates the appropriate statistical procedure to examine the relationship between a continuous variable and a dichotomous categorical variable (two response categories) is an independent-samples t-test. Therefore, we will use this procedure to examine the relationship between the categorical dependent variable *Is the study participant happy?* with the continuous covariate variable *Income Level*. Essentially, this procedure will indicate if there is a statistically significant difference between mean scores reflecting *Income Level* (Response Range: $100k-$500) among study participants that are *Happy* (Yes) or not *Happy* (No). Refer to the corresponding section **4.6.3 Independent-Samples T-Test** in sample study one, for a description of the test assumptions for the independent-samples t-test procedure. To conduct the independent-samples t-test in SPSS go to:

1) **Analyze→ Compare Means→ Independent Samples T-Test**.

2) Move the continuous variable *Income Level* from the left hand column into the cell in the right hand column labeled **Test Variable(s)**.

Figure 5.8 **Define groups**: *Is the study participant Happy?*

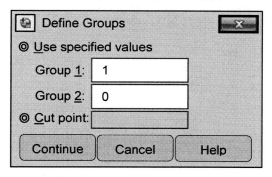

Figure 5.9 **Independent-samples t-test**: *Happiness* by *Income Level*

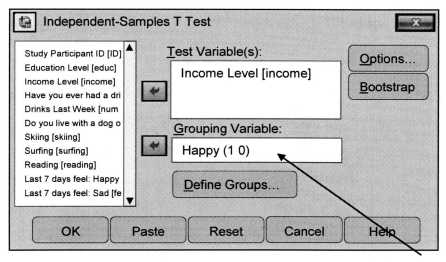

Is the study participant Happy? Yes (=1) or a No (=0)?

3) Move the categorical variable *Is the study participant happy?* from the left hand column into the cell in the right hand column labeled **G̲rouping Variable**.

4) Click the button marked **D̲efine Groups**.

5) In the dialogue box that opens, enter the coded values reflecting *Happy*=Yes (1) and *Happy*=No (0), as presented in **Figure 5.8** where a '1' has been entered into the cell for **Group 1̲**, and a '0' has been entered into the cell for **Group 2̲**.

6) Click **Continue**.

7) The initial dialogue box will look similar to **Figure 5.9**.

8) Click **OK** or **Paste**.

***Interpreting the Statistical Output: Independent-Samples T-Test*. SPSS Output 5.3** presents the statistical output for the independent-samples t-test procedure. In this section we use this statistical output to identify the three dimensions of the relationship between the dependent variable *Happy* and the covariate variable *Income Level*. Specifically, we will present the relationship dimensions of significance, directionality, and magnitude in three separate parts (Part A, Part B, and Part C, respectively).

Part A   Relationship Dimension 1: Significance

Within the statistical output, the second box labeled **Independent-Samples T-Test** presents the level of statistical significance for the procedure. First refer to the column to the left labeled **Levene's Test for Equality of Variances**. The subcolumn beneath labeled **Sig.**, will indicate if the Levene's Test is statistically significant ($p<.05$). If the Levene's test is statistically significant, this indicates that the variances in scores reflecting the continuous variable (*Income Level*) are significantly unequal among the two categorical response groups (*Happy* = Yes, *Happy* = No).

Specifically, if the value in the subcolumn **Sig.** is .05 or below ($p<.05$) the variances are significantly unequal, in which case the values in the second row labeled **Equal Variances not assumed** should be used. In the current analysis we see **Sig.=.000**, indicating the Levene's test is statistically significant ($p<.05$) and we should refer to the bottom row that reads **Equal Variances not assumed**. On the line **Equal Variances not assumed**, under the column labeled **Sig. (2-tailed)**, the level of statistical significance (probability level) for the t-test procedure is reported. We see in this cell **the value .000, indicating that the difference in mean scores is statistically significant ($p<.05$).**

Part B   Relationship Dimension 2: Directionality

The top box in the output (labeled **Group Statistics**) presents the mean *Income Level* scores for the two categorical groups included in the t-test procedure, which are study participants that are *Happy* (Yes or No). Within the column labeled *Is the Study Participant Happy?* the first row labeled *Yes* reflects there are 31 study participants coded as *Happy* (Yes) that have a mean *Income Level* score of 313709.68 (*SD*=101996.14). The second row labeled *No* reflects there are 69 study participants coded as *Happy* (No) with a mean *Income Level* score of 144,927.54 (*SD*=44676.72). Thus, in terms of directionality, the independent-samples t-test indicates that a higher *Income Level* **is associated with a greater likeliness of being** *Happy* **(Yes).**

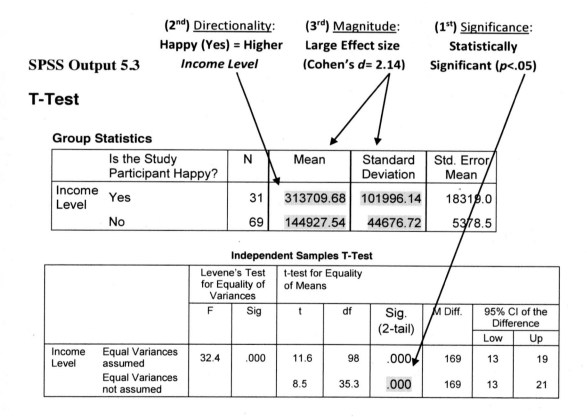

SPSS Output 5.3

(2nd) Directionality: **Happy (Yes) = Higher** *Income Level*

(3rd) Magnitude: **Large Effect size (Cohen's *d*= 2.14)**

(1st) Significance: **Statistically Significant (*p*<.05)**

T-Test

Group Statistics

	Is the Study Participant Happy?	N	Mean	Standard Deviation	Std. Error Mean
Income Level	Yes	31	313709.68	101996.14	18319.0
	No	69	144927.54	44676.72	5378.5

Independent Samples T-Test

		Levene's Test for Equality of Variances		t-test for Equality of Means					
		F	Sig	t	df	Sig. (2-tail)	M Diff.	95% CI of the Difference	
								Low	Up
Income Level	Equal Variances assumed	32.4	.000	11.6	98	.000	169	13	19
	Equal Variances not assumed			8.5	35.3	.000	169	13	21

Part C   Relationship Dimension 3: Magnitude

A common measure of effect size when using an independent-samples t-test is the Cohen's *d*. While the Cohen's *d* is not computed via the independent-samples t-test in SPSS, computing this value is quite straightforward (the formula is presented in **Figure 5.10**). Specifically, to compute Cohen's *d*:

1) **Compute the denominator (the *pooled standard deviation* presented in Figure 5.11):**

    a. Square the **standard deviation** for each group:

          SD of *Happy* (Yes): 101,996.14 = 10403212982.8841

          SD of *Happy* (No): 44,676.72 = 1996009399.3118

    b. Sum those values: 10403212982.8841 + 1996009399.3118 = 12399222382.185941

    c. Divide that summed number by 2: 12399222382.185941/2 = 6199611191.0979705

    d. Take the square root of that number: 78737.6097624

2) **Compute the numerator:** Subtract the mean score of group 1 (313,709.68) from the mean of group 2 (144,927.54), which is = 168,782.14.

3) **Divide the numerator by the denominator:**

4) 168,782.14/78737.6097624 = 2.1436

Figure 5.10 Formula to compute Cohen's *d*

$$d = \frac{Mgroup1 - Mgroup2}{SDpooled}$$

Figure 5.11 The *pooled standard deviation* to compute the Cohen's *d*

$$SDpooled = \sqrt{(SD^2group1 + SD^2group2)/2}$$

An even easier means of computing the Cohen's *d* is to use one of the many calculators available free online (e.g., http://www.uccs.edu/~lbecker/), which will compute this formula for you after you enter the two means and standard deviations for each group in your independent-samples t-test. For example, you will see in **Figure 5.12** we have entered the mean and standard deviation values within our independent-samples t-test and clicked the compute button, which has yielded a Cohen's *d* effect size of 2.14, which matches our manual computation. Recall, a *d*=0.2 is considered a 'small' effect size, 0.5 represents a 'medium' effect size and 0.8 a 'large' effect size. Thus, our *d* of 2.14 reflects a very large effect size between the two variables. Specifically, we would say that the covariate predictor variable *Income Level* has a very large effect on the dependent variable *Happy* (Yes or No).

Figure 5.12 Free online calculator used to compute the Cohen's *d*

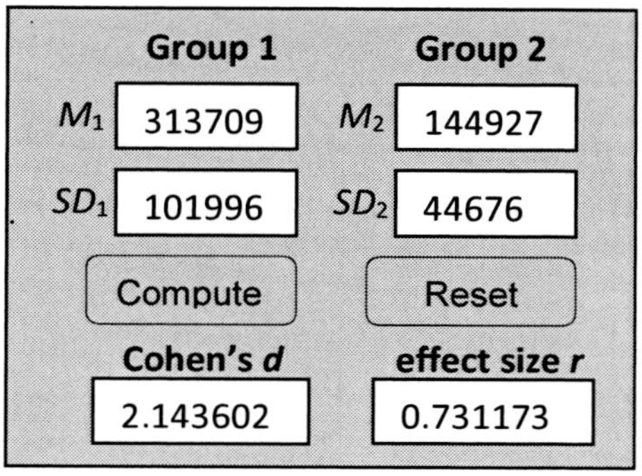

Summary. Overall, the independent-samples t-test indicated a statistically significant relationship between variables (i.e., a difference between mean scores) as the probability level for this test was less than .05 ($p<.05$). Actually, probability is reported as less than .001 ($p<.001$) as the level of probability was .000 (Sig.=.000). In terms of directionality, being coded as *Happy* (Yes) was associated with a higher *Income Level* ($M=313709.68$, $SD=101996.14$) relative not *Happy* (No) ($M=144927.54$, $SD=44676.72$). Lastly, in terms of magnitude, the Cohen's *d* reflected that the covariate predictor variable *Income Level* evidenced a large effect (Cohen's *d* effect size=2.14) upon the dependent variable *Happy* (Yes or No).

5.6.2 Chi-Square

The bivariate test key (**Figure 5.7**) indicates the appropriate statistical procedure to examine the relationship between two categorical variables is a chi-square analysis. Therefore, we will use this procedure to examine the relationships between the dependent variable *Is the study participant Happy?* (Yes or No) with the independent variable *Do You Live with a Dog or a Cat?* and covariate variable *Education Level*. The test assumptions for a chi-square include that each observation (e.g., study participant) is independent of all the others. Specifically, each study participant is within only one group, such as either in the group that lives with a dog or cat, but not both.

Additionally, there is an assumption for a chi-square table larger than 2x2, where no more than 20% of the expected counts are less than 5 and all individual counts are one or greater (Yates, Moore, and McCabe, 1999, p. 734). *Expected counts* are reviewed in this section. Briefly, **each cell** within the crosstabulation table will have a specific expected count of study participants. Furthermore, when a chi-square table is 2x2, there is an assumption that all expected counts are approximately 10 or greater (Cochran, 1952; 1954). When expected counts are somewhat small (<10), the *Fisher exact test for 2x2 contingency table*, might be considered.

Once, these assumptions are considered, the chi-square analysis can be approached. To conduct the chi-square using SPSS, go to:

1) **Analyze→ Descriptive Statistics→ Crosstabs**.

2) In the dialogue box that opens, use the arrow in the upper center to move the predictor variable (e.g., *Education Level*) into the cell on the right hand side marked **R**o**w**(s).

3) Move the dependent variable *Is the study participant Happy?* into the cell marked **C**olumn(s) as presented in **Figure 5.13**.

Figure 5.13   Chi-Square Test: *Education Level* by *Is the study participant Happy?*

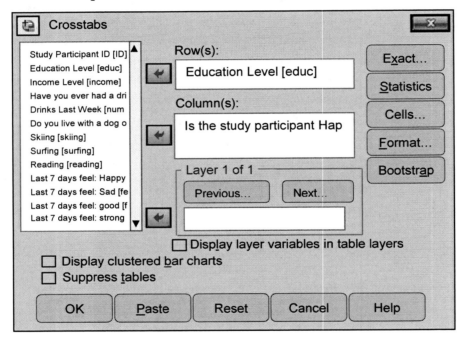

4) Click the button marked **Statistics**.

5) In the dialogue box that opens, check the boxes next to **Chi-Square** and **Phi and Cramer's V** as presented in **Figure 5.14**.

6) Click **Continue**.

7) Click the button marked **Cells**.

8) In the dialogue box that opens, under **Counts**, check **Observed** and **Expected**.

9) Under **Percentages**, check the box next to **Row** as presented in **Figure 5.15**.

10) Click **Continue**.

11) Click **OK** or **Paste**.

Figure 5.14   Activating the Chi-square statistic and Phi and Cramer's V

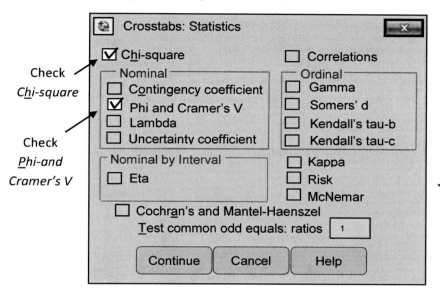

Check
Chi-square

Check
Phi-and
Cramer's V

Figure 5.15   Activating the Observed and Expected Counts and Row Percentages

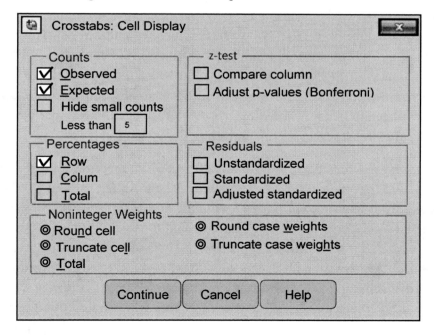

Interpreting the Statistical Output: Chi-Square. **SPSS Output 5.4** presents the statistical output for the chi-square procedure. In this section we use this statistical output to identify the three dimensions of the relationship between the dependent

variable *Happy* (Yes or No) and the independent variable *Do You Live with a Dog or a Cat?* Specifically, we will present the relationship dimensions of significance, directionality, and magnitude in three separate parts (Part A, Part B, and Part C, respectively).

Part A   Relationship Dimension 1: Significance

To observe significance, look at the second box within the **SPSS Output 5.4** labeled **Chi-Square Tests**. Within this box, the first line reads **Pearson Chi-Square**. If the value to the far right on this line, under the column **Asymp. Sig. (2-sided)**, is .05 or below ($p<.05$), then the test indicates a statistically significant relationship between variables. Within the current output, this cell presents **the values .000, which indicates that this test achieves statistical significance**.

Part B   Relationship Dimension 2: Directionality

Next, since the probability level indicates significant differences, we would like to examine what those differences are. Within **SPSS Output 5.4,** refer to the first box labeled **Education Level * Is the study participant happy? Crosstabulation**. Within this box, under the column labeled **Is the study participant happy?**, note the column marked **Yes**. In the intersecting first row for the category *High School/GED*, we see that 9.4% (3 of 32) of study participants are coded as *Happy* (*Happy*=Yes). In the next row, we see that 13.8% (4 of 29) of study participants with a *College* level education are coded as *Happy* (*Happy*=Yes). Finally, the third row reflects 61.5% (24 of 39) of study participants with a *Graduate School* level of education are coded as *Happy* (*Happy*=Yes). Based on these percentages it seems as if a higher level of education is significantly related to being *Happy* (Yes) as the percentage of study participants coded as *Happy* (Yes) increases as levels of education increase.

In addition to these percentages, you can also detect differences through observing the **Count** relative to the **Expected Count** within each cell. Again, refer to the column labeled **Is the study participant happy?**, then the number of study participants in the column labeled **Yes**. You will see an expected count of study participants that describes the number expected if there was no significant association between study variables. Notice in this column through observing *Education Level*, we see the expected count of study participants within the categories *High School/GED* and *College* is 9.9 and 9.0, respectively, while the actual counts for each category is much lower at 3 and 4, respectively. However, the actual count of study participants at the *Graduate School* level of education (actual count = 24) coded as *Happy* (Yes) is approximately twice the expected count (expected count = 12.1).

Thus, we see there are fewer study participants than expected coded as *Happy* (Yes) in the categories *High School/GED* and *College*, but more than expected coded as *Happy* (Yes) at the *Graduate School* level of education. This again suggests there is a greater likelihood of being coded as *Happy* (Yes) for study participants with a higher level of education.

Part C   Relationship Dimension 3: Magnitude

The magnitude of the relationship between two variables in a chi-square analysis where a table is larger than 2 X 2, can be examined using the **Cramer's V** effect size which is computed using the formula presented in **Figure 5.16**:

Figure 5.16   Formula for computing Cramer's V

$$V = \sqrt{\frac{x^2}{N(k-1)}}$$

Within this formula, the N is the total number of observations and k is the number of the rows or columns, whichever smaller, in your crosstabulation table. For example, in terms of our chi-square test results, k would be 2 (we have a 2 X 3 table, with 2 as the smaller number). So our formula would involve:

1) Compute the denominator N $(k-1)$: This is the product of the n (100) multiplied by $k-1$ (where $k-1$ is 2-1=1). Thus, this value is 100 (where 1 X 100 = 100).
2) Next, divide the numerator by the denominator: Specifically, divide the numerator (the value of the Pearson's chi-square), which is 28.01, by the denominator (100), which would be 28.01/100, which is the value 0.2801.
3) Take the square root of the product 0.2801, which is .5293 (effect size estimate).

This value is also presented in our statistical output as a result of checking the box for *Phi and Cramer's V*. Notice within the statistical output, in the third box presented, the value for the **Cramer's V** (i.e., .529) matches our manual computation above. Recall, the effect sizes for **Cramer's V** are 0.1 (small), 0.3 (medium) and large (0.5). Thus, our effect size of .5293 reflects a large effect size, indicating that the Covariate variable *Education Level* has a large effect on the dependent variable *Happy* (Yes or No).

Summary. Overall, the chi-square analysis indicated a statistically significant relationship between variables as the probability level for this test was less than .05 ($p<.05$). Actually, probability is reported as less than .001 ($p<.001$) as the level of probability was .000 (Sig.=.000). In terms of directionality, there seemed to be an association between a higher *Education Level* and a greater likelihood of being *Happy*

(Yes), as a significantly higher percentage of study participants with a *Graduate School* level of education (61.5%; 24 of 39) were *Happy* (Yes), relative to those with a *High School/GED* (9.4%; 3 of 32) and *College* (13.8%; 4 of 29) level of education. Lastly, in terms of magnitude, the Cramer's V reflected that the covariate predictor variable *Education Level* evidenced a large effect (Cramer's V = 0.53) upon the dependent variable *Happy* (Yes or No).

SPSS Output 5.4

Crosstabs

Education Level * Is the study participant happy? Crosstabulation

			Is the study participant happy? No	Is the study participant happy? Yes	Total
Education Level	High School/GED	Count	29	3	32
		Expected Count	22.1	9.9	32.0
		% within Education Level	90.6%	9.4%	100.0%
	College	Count	25	4	29
		Expected Count	20.0	9.0	29.0
		% within Education Level	86.2%	13.8%	100.0%
	Graduate School	Count	15	24	39
		Expected Count	26.9	12.1	39.0
		% within Education Level	38.5	61.5%	100.0%
Total		Count	69	31	100
		Expected Count	69.0	31.0	100.0
		% within Education Level	69.0%	31.0%	100.0%

Chi-Square Tests

	Value	df	Asymp. Sig. (2-sided)
Pearson Chi-square	28.01	2	.000
Likelihood Ratio	28.6	2	.000
Line-by-Line Assoc.	23..2	1	.000
N of Valid Cases	100		

(2ⁿᵈ) Directionality:
Higher Education Level
= Higher % *Happy*

(1ˢᵗ) Significance:
Statistically significant (*p*<.05)

Symmetric Measures

		Value	Approx. Sig.
Norm by Norm	Phi	.529	
	Cramer's V	.529	.000
	N of Valid Cases	100	.000

(3ʳᵈ) Magnitude:
Large Effect size
(Cramer's V = 0.53)

Conduct the Second Chi-Square Test Using the Independent Variable:

Repeat the test using the Independent Variable. Next, repeat this procedure, but remove the covariate variable *Education Level* and enter the independent variable, *Do You Live with a Dog or a Cat?* in the box that the covariate variable *Education Level* was formerly. This will indicate if there is a significant association between the independent variable *Do You Live with a Dog or a Cat?* and the dependent variable *Is the study participant Happy?*

Interpreting the Statistical Output: Chi-Square. **SPSS Output 5.5** presents the statistical output for the chi-square procedure. In this section we use this statistical output to identify the three dimensions of the relationship between the dependent variable *Happy* (Yes or No) and the independent variable *Do You Live with a Dog or a Cat?* Specifically, we will present the relationship dimensions of significance, directionality, and magnitude in three separate parts (Part A, Part B, and Part C, respectively).

Part A   Relationship Dimension 1: Significance

To observe significance, look at the second box in the **SPSS Output 5.5** labeled **Chi-Square Tests**. Within this box, the first line reads **Pearson Chi-Square**. If the value to the far right on this line, under the column **Asymp. Sig. (2-sided)**, is .05 or below ($p<.05$), then the test indicates a statistically significant relationship between variables. Within **SPSS Output 5.5**, this cell presents **the value .000, which indicates that this test achieves statistical significance**.

Part B   Relationship Dimension 2: Directionality

Next, refer to the first box labeled **Do You Live with a Dog or a Cat? * Is the study participant happy? Crosstabulation**. Within this box, under the column labeled **Is the study participant happy?**, note the column marked **Yes**. In the intersecting first row for the variable *Do You Live with a Dog or a Cat?* we see that 48.9% (23 of 47) of study participants that live with a dog are coded as *Happy* (*Happy*=Yes). Additionally, we see that 15.1% (8 of 53) of study participants that live with a cat are *Happy* (*Happy*=Yes). Regarding the **Count** relative to the **Expected Count**, we see the expected count of study participants coded as *Happy* that live with a dog is 14.6, but the actual count is much higher at 23. Furthermore, the expected count of study participants coded as *Happy* (Yes) that live with a cat is 16.4, but the actual count is half that value (8). Thus, there are more than expected study participants that live with a dog coded as *Happy* (Yes) and less than expected that live with a cat coded as *Happy* (Yes), suggesting a greater likelihood of being *Happy* (Yes) when living with a dog.

SPSS Output 5.5

Crosstabs

Do You Live with a Dog or a Cat?* Is the study participant Happy? Crosstabulation

			Is the study participant happy? No	Is the study participant happy? Yes	Total
Do You Live with a Dog or a Cat?	Dog	Count	24	23	47
		Expected Count	32.4	14.6	47.0
		% within Livedogcat	51.1%	48.9%	100.0%
	Cat	Count	45	8	53
		Expected Count	36.6	16.4	53.0
		% within Livedogcat	84.9	15.1%	100.0%
Total		Count	69	31	100
		Expected Count	69.0	31.0	100.0
		% within Education Level	69.0%	31.0%	100.0%

Chi-Square Tests

	Value	df	Asymp. Sig. (2-sided)
Pearson Chi-Square	13.337	1	.000
Continuity Correction	11.8		.001
Likelihood Ratio	13.7	1	.000
Fisher's Exact Test			
Linear-by-Linear	13.2	1	.000
N of Valid Cases	100		

(2ⁿᵈ) Directionality:
Lives with Dog =
Higher % *Happy*

(1ˢᵗ) Significance:
Statistically significant
($p<.05$)

Symmetric Measures

		Value	Approx. Sig.
Nominal by Nominal	Phi	.365	
	Cramer's V	.365	.000
	N of Valid Cases	100	.000

(3ʳᵈ) Magnitude:
Medium effect size
(Phi = 0.37)

Part C   Relationship Dimension 3: Magnitude

To test the effect size for a chi-square analysis of variables presented in a 2 X 2 table, we can determine the value of Phi (φ), which is computed by taking the square root of the product of the value of Chi/the number (n) of observations (see **Figure 5.17**).

Figure 5.17   Formula for computing Phi

$$\varphi = \sqrt{\frac{x^2}{n}}$$

Thus, within the current example, we would:

1) Take the value of Chi (the Pearson Chi-Square value = 13.337) and divide that value by the number of observations (100), which would produce the value .13337 (13.337/100=.13337).

2) We would then take the square root of that product (.13337), which would be .3652, which would reflect the Phi effect size relationship.

Note, this value of **Phi** is also presented in the third box within our **SPSS Output 5.5** as a result of checking the box next to the term *Phi and Cramer's V*. This estimate of Phi within the statistical output (.365) matches our manual computation. Recall, the effect sizes for Phi are 0.1 (small), 0.3 (medium) and large (0.5). Thus, our effect size of .3652 reflects a medium effect size, indicating that the independent variable *Do You Live with a Dog or a Cat?* has a medium size effect on the dependent variable *Happy* (Yes or No).

Summary. Overall, the chi-square analysis indicated a statistically significant relationship between variables as the probability level for this test was less than .05 ($p<.05$). Actually, probability is reported as less than .001 ($p<.001$) as the level of probability was .000 (Sig.=.000). In terms of directionality, there seems to be an association between living with a dog and a greater likelihood of being *Happy* (Yes), as a significantly higher percentage of study participants that *live with a dog* (48.9%; 23 of 47) were coded as *Happy* (Yes), relative to those that *live with a cat* (15.1%; 8 of 53). Lastly, in terms of magnitude, the Cramer's V reflected that the independent variable *Do You Live with a Dog or a Cat?* evidenced a medium size effect (Phi = 0.37) upon the dependent variable *Happy* (Yes or No).

5.6.3   What Did Bivariate Analysis Tell Us?

Our primary goal in bivariate analysis is to identify which predictor variables (covariate/independent variables) have a statistically significant relationship with the dependent variable so these predictors may be included in **Step 6: Multivariate Analysis**. Although we examined all three dimensions of the relationships between variables (significance, directionality, and magnitude) in our bivariate testing, below we will only review the dimension of statistical significance. **Figure 5.18** presents the results of the three bivariate tests used in sample study two examining which predictor variables have a statistically significant association with the dichotomous categorical variable *Is the study participant Happy?* (Yes or No).

Notice that we have entered an arrow connecting Covariate Variable 1: *Education Level* with the dependent variable, which indicates a chi-square analysis revealed a statistically significant relationship ($p<.05$). Likewise, there are arrows reflecting statistically significant relationships between Covariate Variable 2: *Income Level* and the Independent Variable: *Do You Live with a Dog or Cat?* with the dependent variable.

Based on these our findings, we will include all three predictor variables in the next step, multivariate analysis, as each is related to the dependent variable at a statistically significant level. Please note, at this step we can only see which predictors are significantly related to the dependent variable, but we cannot see which variable is the strongest predictor of being *Happy* (Yes or No). That dimension will be revealed in multivariate analysis!

Figure 5.18   What Bivariate Analysis Tells Us for Sample Study Two?

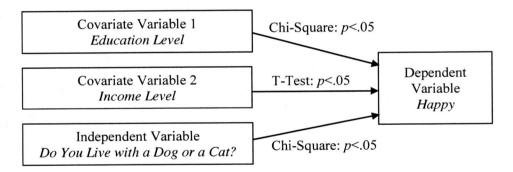

5.7   Step 6: Multivariate Analysis

In this section, we will employ multivariate analysis to identify the strongest predictor of being *Happy* (Happy=Yes) among the predictor variables (covariate and independent variables) that are significantly related to the dependent variable at the bivariate level ($p<.05$). Recall, in the prior section we identified that both covariate predictor variables *Education Level* and *Income Level*, as well as the independent variable *Do You Live with a Dog or a Cat?*, are significantly related to the dependent variable at the bivariate level and will be included in the multivariate analysis.

Next, the correct multivariate test must be selected. **Figure 5.19** presents the **Multivariate Test Key with the variables for sample study two plugged in**. This key describes the appropriate multivariate procedure when examining a dichotomous categorical dependent variable, such as *Happy* (Yes or No), is a binary logistic regression. Binary logistic regression can be used to predict the value of a dichotomous dependent variable based on the value of two or more other predictor variables, while observing the relative contribution of each.

Figure 5.19 Multivariate Test Key with the study variable plugged in

Dependent Variable	Regression to Use
Categorical (≥ 3 categories)	Multinomial Logistic (Not Presented here)
Categorical (2 categories)	Binary Logistic
Categorical (2 categories)	Binary Logistic
Continuous	Linear

Before performing the binary logistic regression, it is important to confirm that all necessary test assumptions have been addressed. For example, **Section 5.4.2 Checks of Test Assumptions** described the need to assess that no significant **multicollinearity** exists between predictors for this regression model, which was substantiated. A second assumption that must be met is that all predictor variables in the regression model are at the continuous or dichotomous level of measurement. Since the categorical covariate predictor variable *Education Level* has a three category response, the categories within that variable must be dummy-coded (as presented in **Section 5.7.1**) prior to use in the multivariate model. After the process of dummy-coding has been addressed, the data will be appropriate for use in the binary logistic regression model.

5.7.1 Dummy-Coding Variables

Please see the corresponding section in sample study one **4.7.1 Dummy-Coding Variables** for a full description of the process needed to dummy-code a predictor variable. In short, to use a categorical variable with more than two categories in a regression model, the variable must be dummy-coded. This involves making each response category within the original categorical variable, into a separate variable with a two category response (most often Yes or No). For example, in the current analysis the covariate predictor variable *Education Level* is a categorical variable with more than two response categories. Specifically, there are three response categories (High School/GED, College, Graduate School) within the variable *Education Level*. Thus, to use the variable *Education Level* in a regression model, we must dummy-code each response category toward creating three separate variables with a Yes or No response:

1) High School/GED (Yes or No)

2) College (Yes or No)

3) Graduate School (Yes or No)

Next, one of the three dummy-coded categories must be specified as the **reference group**. In sample study two, we will select the category *High School/GED* (Yes or No) to be used as the reference group (as we did in sample study one). Recall, the group selected as the reference group is the predictor left out of the regression model when the analysis is conducted. Again, please see the corresponding section **4.7.1 Dummy-Coding Variables** in sample one for a further explanation. This process of recoding the variable *Education Level* into three separate dummy-coded variables has already been completed in sample study one. Subsequently, we will plan on using these three dummy-coded variables in analysis and move on to the process of conducting the binary logistic regression model.

5.7.2　Conducting Binary Logistic Regression

Our goal in multivariate testing is to test the association between multiple predictor variables with the dependent variable *Happy* (Yes or No). To conduct a Binary Regression model in SPSS go to:

1) **Analyze→ Regression→ Binary Logistic**.

2) In the dialogue box that opens, move the dependent variable *Happy* (Yes or No) from the left hand column into the right hand column into the cell labeled **Dependent**.

3) Enter the two dummy-coded variables *EducationGraduateSchool* and *EducationCollege* (reflecting *Education Level*) into the cell labeled **Covariates**. Recall, we are leaving the third dummy-coded variable *EducationHighSchoolGED* out of the model to serve as the reference group.

4) Enter the covariate predictor variable *Income Level* into the cell labeled **Covariates**.

5) Enter the independent variable *Do You Live with a Dog or Cat?* into the cell labeled **Covariates**, as presented in **Figure 5.20**.

6) Click the button labeled **Options**.

7) In the dialogue box that opens, under the section **Statistics and Plots**, check the box next to **CI for exp(B)** (as presented in **Figure 5.21**), which will produce the 95% confidence interval.

8) Click **Continue**.

9) Click **OK** or **Paste**.

Figure 5.20   Conducting binary logistic regression

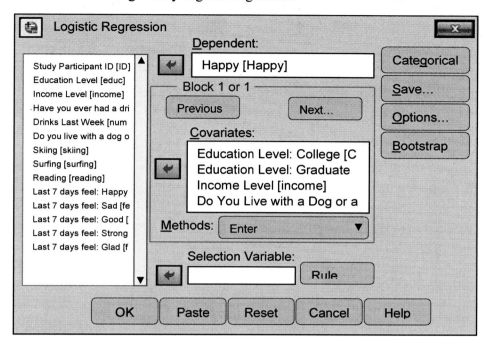

Figure 5.21   Activating the 95% confidence interval in binary logistic regression

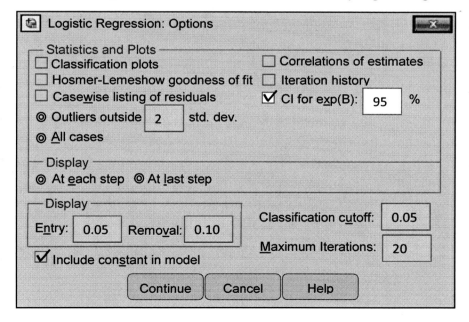

Interpreting the Statistical Output: Binary Logistic Regression Analysis. **SPSS Output 5.6** presents the statistical output for the **Binary Logistic Regression Analysis** procedure. In this section we use this statistical output to identify the three dimensions of the relationship between the dependent variable *Happy* and predictor variables (covariate and independent variables) in the context of the full model. Specifically, we will present the relationship dimensions of significance, directionality, and magnitude in three separate parts (Part A, Part B, and Part C, respectively).

Part A   Relationship Dimension 1: Significance

Significance of the Overall Model. Within **SPSS Output 5.6**, the first box (labeled **Omnibus Tests of Model Coefficients**) presents the level of statistical significance for the overall binary logistic regression model. Specifically, in the bottom line (labeled **Model**) we see the value of the chi-square statistic (58.555) and level of statistical significance (**Sig.**=.000) for the regression model. If the probability level is greater than .05, there is no need to evaluate the rest of the model as the findings are regarded as occurring by chance. However, we see in this output that the level of probability for the binary logistic regression model for sample study two is .000. Thus, we will review the model further. While assessing the full model, you might also note the percentage of cases categorized correctly, which is 85%. This is another common metric reported for the logistic regression model.

Significance of the Individual Predictor Variables. Next, within the box labeled **Variables in the Equation (A)**, note the last column labeled **Sig.**, which presents the level of statistical significance between each individual predictor and the dependent variable. The top two rows of this column describe the dummy-coded variable *EducationCollege* has a probability level of .732, while *EducationGraduateSchool* has a probability level of .000. Remember these results *are in reference* to the reference group, which is *EducationHighSchoolGED*. Therefore, these probability values indicate that the *EducationCollege* group (p=.732) is not significantly more likely to be *Happy* (Yes) relative to the *EducationHighSchoolGED* group. However, the *EducationGraduateSchool* group is significantly (p<.05) more likely to be *Happy* (Yes) relative to the *EducationHighSchoolGED* group as the **Sig.** is **.000** (p<001).

The next line within the **Sig.** column reflects the relationship between the continuous covariate variable *Income Level* and the dependent variable *Happy* (Yes or No) is statistically significant, as the value (**Sig.=.003**) is less than .05 (p<.05). The fourth line in this column indicates the relationship between the independent variable *Do You Live with a Dog or a Cat?* and the dependent variable *Happy* (Yes or No) is statistically significant as the value (**Sig.=.002**) is below .05 (p<.05).

Part B   Relationship Dimension 2: Directionality

The Beta. Next, to identify directionality, again refer to the column labeled **B** within the table titled **Variables in the Equation (A)**, which presents the beta coefficient. Recall, when the beta is preceded by a minus, a negative relationship is indicated, while the absence of a minus sign reflects a positive relationship. For example, the second row within this column (we do not consider the first line as *p*>.05) indicates a beta of 3.17, which indicates a positive relationship where the *EducationGraduateSchool* group is significantly more likely (positive indicates more likely) to be *Happy* (*Happy*=Yes) in reference to the *EducationHighSchoolGED* group. The third line presenting *Income Level*, evidences a positive beta of .04, which indicates as *Income Level* increases the study participant is more likely to be *Happy* (Yes).

Lastly, the fourth line presents the directionality between the independent variable *Do You Live with a Dog or a Cat?* and the dependent variable *Happy* (Yes or No). Observe the beta (B= -2.26) is negative, indicating a negative relationship reflecting as the code for the variable *Do You Live with a Dog or a Cat?* increases from 1 to 2, (1=dog, 2=cat), the code for the dependent variable decreases from 1 to 0, which signals a movement from *Happy* (1=Yes) to not *Happy* (0=No). Thus, there is a significant negative relationship between living with a cat (Cat=2) and being Happy (*Happy*=1).

Part C   Relationship Dimension 3: Magnitude

As previously mentioned in logistic regression, the effect size between relationships is expressed via the odds ratio. Although estimates of what constitutes a small, medium, and large effect size for an odds ratio often varies, a generally accepted convention is **small = 1.44, medium = 2.47, and large = 4.25**. Within the statistical output the odds ratio is reported in the column labeled **Exp(B)**. The confidence interval for each odds ratio is presented in the two columns (lower and upper bound estimates) to the right.

The 95% confidence interval. The 95% confidence interval is our way of acknowledging that the odds ratio generated in statistical analysis is not likely the actual odds ratio that exists in the target population as a whole. However, we are suggesting that we are 95% confident that the true odds ratio within the target population exists between the values presented in the 95% confidence interval. The wider the 95% confidence interval, the less stability is indicated in the finding.

In sample study two, the odds ratios can be found within the box labeled **Variables in Equation (B)** in the column marked **Exp(B)**. Within this column, on the second line, we see the odd ratio for the dummy-coded variable *EducationGraduateSchool* is 23.85. This value suggests that study participants in the group *EducationGraduateSchool* are

23.85 times more likely to be *Happy* (Yes), in reference to study participants in the *EducationHighSchoolGED* group (the reference group). Thus, a great magnitude is indicated as there is a large effect size between the variable *Education Level* and *Happy* (Yes or No). However, the 95% confidence interval for this odds ratio is extremely wide (95% CI = 4.19-135.67), suggesting the finding is unstable.

Next, on the third line, within the column **Exp(B)**, we see an odds ratio of 1.04 for the covariate variable *Income Level*. However, it is important to realize that an odds ratio in a logistic regression model is *not interpretable for a continuous predictor variable in the same way as a categorical predictor variable*. Specifically, when a predictor is categorical (e.g., *Education Level*) in binary logistic regression, it is appropriate to report a greater likelihood of experiencing the event. For example, we could state that those with a *Graduate School* level education are 23.85 times more likely to be *Happy* relative to the reference group.

However, when the predictor variable is continuous, such as *Income Level*, the odds ratio *is not interpretable in the same way, where one might state an outcome is more or less likely to occur*. Specifically, when the predictor is continuous, the odds ratio may be reported, but not phrased in terms of the event being more or less likely to happen. For example, in the case of the covariate predictor variable *Income level*, we might choose to simply report that a higher *Income Level* is significantly associated with experiencing the event, being *Happy* (Yes), as well as the odds ratio (OR=1.04). However, we might refrain from making a statement regarding how many times more or less likely a study participant is to experience the outcome.

The fourth line within the column labeled **Exp(B)**, presents the odds ratio between the independent variable *Do You Live with a Dog or Cat?* and the dependent variable *Happy* (Yes or No), which is .105. This odds ratio is different than the other odds ratios, as the value is below 1.00 (OR=.105). *An odds ratio below 1.0 suggests a negative association with the outcome reflected in the dependent variable*. For example, an odds ratio below 1.0 suggests a negative relationship with experiencing the outcome of being *Happy* (*Happy*=Yes). However, an odds ratio below is not easily interpreted, so many times these odds ratios are converted to a number greater than 1.0 to better communicate the magnitude of the variable relationship.

Converting a Negative Odds Ratio Manually. To convert an odds ratio below 1.0 manually simply divide 1.0 by the odds ratio. For example, if we divide 1 by the odds ratio .105, the resulting number is 9.5 (1.0/.105=9.5). This suggests that study participants that live with a dog are 9.5 times more likely to be coded as *Happy* (*Happy*=Yes), relative to study participants that live with a cat.

SPSS Output 5.6

Logistic Regression

Omnibus Tests of Model Coefficients

		Chi-square	df	Sig.
Step 1	Step	58.555	4	.000
	Block	58.555	4	.000
	Model	58.555	4	.000

(1st) Significance (model): Statistically significant (p<.05)

Classification Table[a]

Observed		Predicted		
		Is the study participant happy?		Percent Correct
		No	Yes	
Is the study part-icipant happy?	Yes	62	7	89.9
	No	8	23	74.2
Overall percent				85.0

Percentage of cases categorized correctly

(1st) Significance (predictors): Statistically significant (p<.05)

(2nd) Directionality: Beta (B) indicates all positive relationships except for the IV

Variables in Equation (A)

		B	S.E.	Wald	df	Sig.
Step 1	EducationCollege	.33	.95	.12	1	.732
	EducationGraduateSchool	3.17	.88	12.78	1	.000
	Income Level	.04	.01	9.00	1	.003
	Livedogcat (dog=1, cat=2)	-2.26	.74	9.32	1	.002
	Constant	2.40	1.19	4.09	1	.043

(3rd) Magnitude: Odds ratio effect sizes are large

Variables in Equation (B)

		Exp (B)	95% Confidence Interval of the Difference	
			Lower	Upper
Step 1	EducationCollege	1.39	.21	9.01
	EducationGraduateSchool	23.85	4.19	135.67
	Income Level	1.04	1.01	3.17
	Livedogcat (dog=1, cat=2)	.105	.03	.45
	Constant	11.05		

Converting an Odds Ratio by Switching the Independent Variable Response Category Coding. You can also convert a negative odds ratio by changing the numbers used to code the response categories in the SPSS database. For example, **SPSS Output 5.7** presents the binary logistic regression analysis model for sample study 2, with the coding reversed for the independent variable *Do You Live with a Dog or a Cat?* where cat is coded as 1 and dog is coded as 2 (instead of the original coding of dog = 1 and cat = 2). Under the column **Exp(B)**, on the line for the variable *Livedogcat*, the odds ratio is now expressed as a number greater than 1.0, which is a more interpretable 9.55.

Furthermore, the 95% confidence interval is also more easily interpreted when the odds ratio is greater than 1.0. For example, the 95% confidence interval for the variable *Livedogcat* is now 2.24-40.67. Thus, the data indicate that study participants who live with a dog are almost ten times (OR=9.55) more likely to be *Happy* (Yes) relatively to study participants that live with a cat. However, we should also acknowledge that the 95% confidence interval for the odds ratio is very wide (95% CI = 2.24-40.67), which suggests that the findings are not exceptionally definitive.

SPSS Output 5.7

		B	S.E.	Wald	df	Sig.
Step 1	EducationCollege	.33	.95	.12	1	.732
	EducationGraduateSchool	3.17	.88	12.78	1	.000
	Income Level	.04	.01	9.00	1	.003
	Livedogcat (**cat=1, dog=2**)	2.257	.739	9.324	1	.002
	Constant	2.40	1.19	4.09	1	.043

Coding is reversed:
cat=1 and dog=2

		Exp (B)	95% Confidence Interval of the Difference	
			Lower	Upper
Step 1	EducationCollege	1.39	.21	9.01
	EducationGraduateSchool	23.85	4.19	135.67
	Income Level	1.04	1.01	3.17
	Livedogcat (**cat=1, dog=2**)	9.553	2.24	40.67
	Constant	11.05		

OR is above 1.0 and more
easily interpreted

5.7.3   What Did Multivariate Analysis Tell Us?

Figure 5.22 describes the findings within the multivariate analysis for sample study two. Before observing the figure, it is important to note that data indicated the overall logistic regression model was statistically significant. Each predictor variable was significantly related to the dependent variable, as indicated within this figure by the probability notation on each of the arrows. Specifically, study participants with a *Graduate School* level education were significantly more likely to be Happy (Yes), relative to the *High School/GED* group, but not the group with a *College* level education. Furthermore, an increase in *Income Level* was significantly associated with being coded as *Happy* (Yes). Lastly, study participants that lived with a dog were significantly more likely to be *Happy* (Yes) relative to those that live with a cat.

In terms of the magnitude of the relationships, we noted large odds ratio effect sizes. For example, study participants with a *Graduate School* level education were described as being 23.85 times more likely to be *Happy* (Yes), relative to the *High School/GED* group. Additionally, study participants that lived with a dog were almost ten times (9.55) more likely to be *Happy* (Yes) relative to those that live with a cat.

However, as we noted, the 95% confidence intervals were rather wide for these estimates, reflecting that these odds ratios might be rather unstable numbers. The confidence interval for the odds ratio regarding levels of education was especially wide at 4.19-135.67. This wide interval might suggest that although the odds ratio for this group (OR=23.85) is larger than the independent variable (OR=9.55), the relationship reflected between the independent and dependent variable might be more believable.

Figure 5.22   What multivariate analysis told us for sample study two

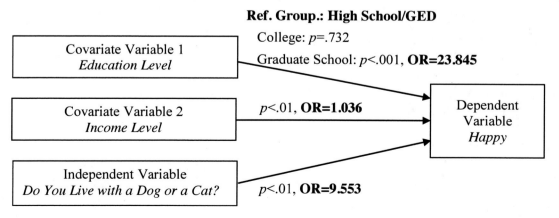

5.8   Step 7: Write-up & Report

Please see the corresponding **Section 4.8** in sample study one for a detailed presentation of the elements that should be presented in a research report, such as a peer-reviewed journal article. In this section (**5.8**) we will only present the features of the paper that would differ when a binary logistic regression model is used as in sample study two, instead of a multiple linear regression model as presented in the sample manuscript included in sample study one.

5.8.1   Changes to the Body of the Paper

In the sample manuscript presented in **Section 5.8.2** for sample study two, the **Abstract** has several changes relative to sample study one. For example, the outcome under examination is now *Happy* (Yes or No), rather than *Happiness*. Furthermore, instead of the multiple linear regression model, we refer to the binary logistic regression model used in sample study two, along with an associated odds ratios.

You will see that the primary changes to the **Introduction** and **Literature Review** sections for the sample manuscript for sample study two, is changing the reference to the outcome variable from *levels of Happiness*, to *a greater likelihood of being Happy*. This is because sample study one examines a continuous dependent variable (*Happiness*) and sample study two examines a dichotomous categorical variable (*Happy*, Yes or No).

Regarding the **Methods**, the sections on Study Participant Recruitment, Research Design, and Sample would remain the same relative to the Write-up and Report presented in sample study one. Also, the Measurement instruments section would remain the same with one exception. For sample study two we need to slightly revise the description to describe how the dependent variable *Happy* (Yes or No) is measured. You will see within the sample manuscript the title for this section has been changed from *Happiness* (from sample study one) to *Happy* (Yes or No). You will also see that we added a description of how the cutoff point to determine which scores can be categorized as *Happy* (Yes or No).

The Data analysis plan has also been modified slightly. The same three phases are presented, **Step 4: Univariate Analysis**, **Step 5: Bivariate Analysis**, and **Step 6: Multivariate Analysis**, along with a subsequent mention of **Step 3: Checks of Data Integrity**. The modification occurs in changing the dependent variable to a dichotomous categorical outcome, as well as reducing the number of checks of tests assumptions that must be made.

You will see the **Results** section is laid out in the same way as the sample manuscript in sample study one. The modification in the manuscript for sample study two is the statistical tests applied in the analysis. Otherwise, the format mirrors the corresponding section in sample study one. Lastly, the **Discussion** and **Conclusion** sections for sample study two have the same formats of the first study, but reflect the specific findings of the second analysis.

5.8.2 The Sample Manuscript for Sample Study Two

THE RELATIONSHIP BETWEEN BEING HAPPY AND LIVING WITH A DOG OR CAT

Abstract

The current study examines if participants living with a dog are significantly more likely to be happy (Yes) relative to those living with a cat. The study employs a cross-sectional design and involves a convenience sample of residents living in an apartment complex in a major Northeastern city. A multivariate binary logistic regression model indicated that study participants that lived with dogs were almost ten times (OR=9.55) more likely to be happy relative to those that lived with cats. Higher levels of education and income were also associated with a greater likelihood of being happy. Future research would benefit from a mixed-methods design that might include a qualitative piece capturing participant perspectives regarding why dog ownership might be associated with a greater likelihood of being happy relative to cat ownership.

Keywords: Happiness, Pets, Dogs, Cats

INTRODUCTION

(An Overview of the Central Issue and Reason for the Study)

The study of being *Happy* is a well-researched area that has been related to several outcomes, including education level (*citation*), income level (*citation*) and several other factors (*citation*). However, being happy has not been examined in relation to living with a dog or cat. Therefore, the current study will examine if study participants that live with a dog are more likely to be happy (Yes/No) relative to those that live with a cat.

LITERATURE REVIEW

(**1st** *Discuss the Dependent Variable Construct*)

Research examining the factors related to being happy is complimented by a full body of work examining related factors such as general life satisfaction (*citation*), elevated mood (*citation*),….However, quantitative studies that examine the outcome of being happy are clearly identifiable through assessing that particular emotion with a specialized battery of measurement tools (*citation*)…

(*Discuss Outcomes Linked to the Dependent Variable,*

but not the Independent Variable Yet)

One of the most consistent predictors of being happy is income level (*citation*). Specifically, numerous studies have linked higher levels of income to a greater likelihood of being happy (*citation*)…

Being happy has also been linked to higher educational attainment (*citation*). Smith and colleagues (*citation*) identified that….

(**2nd** *Discuss the Independent Variable Construct*)

Pet ownership has been linked to several outcomes (*citation*). For example, elderly residents of a specialized living facility evidenced a greater likelihood of life satisfaction when a pet resided with them in their apartment (*citation*). Similar findings are reflected with other age groups…..

Although pet ownership has been linked to several outcomes, research has not been conducted that examines how these outcomes differ based upon pet type. Specifically, data have not been gathered that reflect if outcomes might differ when individuals reside with a dog, cat,…

(**3rd** *Discuss the Link between the Dependent and Independent Variables*)

Research has not been conducted that examines if being happy is more or less likely relative to the type of pet an individual selects. However, there is evidence to suggest such a relationship might exist, where a group of pet owners with pets that demanded more care, such as cats, evidenced higher levels of stress relative to those who owned more independent animals, such as dogs (*citation*). Specifically,…

(**4th** *In the Final Paragraph Before the Methods, Enter the Study Map Information*)

The current study will examine if people who live with dogs are more likely to be happy relative to people who live with cats. The analysis will also include several covariate variables that have been shown to impact being happy, such as education and income levels. This analysis represents the first attempt at gathering preliminary data that might suggest the impact of pet ownership on being happy might vary by pet type.

METHODS

Measurement instruments

Happy (Yes or No). Being Happy (Yes or No) will be measured via a cutoff score after the computation of the 5-item scale created for this study, which reflects the construct *Happiness*. Questions on the 5-item measurement instrument are measured along a Likert-type scale using a continuum of 1 (Strongly Disagree) to 5 (Strongly Agree). The scale is computed through taking the mean score of valid responses, resulting in a potential range of scores of 1-5, with higher scores indicating a greater degree of *Happiness*. A reliability analysis indicated the level of scale internal consistency was acceptable for the sample (Cronbach's alpha=.89).

The cutoff score reflecting being *Happy* (Yes/No) is computed through identifying scores ½ a standard deviation above the mean score for the sample. The mean score reflecting *Happiness* is 2.73 and one standard deviation is .76. Thus, the mean score of 2.73 plus one half a standard deviation .38 (1 SD= .76 divided by 2=.38) is 3.11. Subsequently, *Happiness* scores greater than 3.11 (3.12 to 5.00) are coded as *Happy* (*Happy*=Yes), while the remaining scores (1.00 to 3.11) are coded as not *Happy* (*Happy*=No).

Data analysis plan

Data analysis will be conducted in three phases. First, all data will be analyzed descriptively via univariate analysis. Second, the relationship between the dependent variable *Happy* with the independent (*Do you live with a dog or cat?*) and covariate variables (*Education Level, Income Level*) will be examined using bivariate analysis. Predictor variables (independent/covariate variables) associated with the dependent variable at a statistically significant level ($p<.05$) will be entered in the final multivariate model. Third, a binary logistic regression model will be used to model the dependent variable *Happy* (Yes/No) as a function of all predictor variables included in the multivariate model.

Prior to data analysis, an examination of test assumptions indicated multicollinearity would not present a significant problem in analysis. A power analysis indicated 100 study participants would provide sufficient statistical power to detect a medium size effect between the independent and dependent variable, which reflected the current sample of 100 study participants should provide sufficient statistical power for this analysis.

Data were initially collected from 105 study participants. Of these, 10 evidenced missing data values. All missing data values pertained to the five items that compose the 5-item scale measuring *Happiness*. Five of these study participants were excluded from analysis because each failed to provide valid responses to at least 80% of items that compose the 5-item *Happiness* scale, which left a final sample of 100 study participants for analysis. The other five study participants with missing data values, provided valid responses for 80% of scale items (4 of the 5 scale items) and

were included in the analysis through using the mean score of the valid responses.

Bivariate analysis indicated data appeared to be missing at random, as study

participants with and without missing data values did not differ significantly by key

characteristics, including education level, X (2)=.41, p=.82, and income level,

t(103)=.36, p=72.

RESULTS

Descriptive Analysis

Table 1 presents a descriptive analysis of the categorical study variables

Education level, Do you live with a dog or cat? and *Is the study participant happy?*

Data indicated that the sample of 100 study participants were evenly divided by

education level, where about one-third (32%; n=32) indicated a highest level of

education of High School/GED, while 29% (n=29) indicated a college (4 year degree)

level and 39% (n=39) indicated a graduate school level education. Approximately half

of study participants reported living with a dog (n=47; 47%) or cat (n=53; 53%).

Additionally, the scores of approximately one-third (n=31; 31%) of study participants

reflected being happy (Yes), while the remaining study participants (n=69; 69%) were

coded as not happy. Table 2 presents a descriptive analysis of the continuous study

variables income level and happiness. Data indicated that the average score reflecting

happiness as 2.73 (SD=.76; MIN/MAX=1.20-4.40) and income level as $197, 250.00

(SD=103,336.79; MIN/MAX=100,000-495,000).

Bivariate Analysis

Table 3 presents a bivariate analysis of the variable income level by the dependent variable Is the study participant happy? An independent-samples t-test reflected that the mean score reflecting income level was significantly higher among study participants coded as happy (M=$313,709.68, SD=101996.14) relative to study participants coded as not happy (M=$144,927.54, SD=44676.72), $t(98)$=11.6, p<.001.

Table 4 presents a bivariate analysis of the variables education level and Do you live with a dog or cat? with the dependent variable Is the study participant happy? Chi-square analysis indicated that higher levels of education were associated with higher percentage of study participants coded as happy, $X(2)$=28.01, p<.01. Specifically, 9.4% (n=3) of study participants with a high school/GED level of education were happy. Additionally, 13.8% (n=4) of study participants with a college level of education were coded as happy. However, 61.5% (n=24) of study participants with a graduate school education level were coded as happy. Chi-square analysis also indicated that a significantly higher percentage of study participants that lived with a dog were happy (48.9%; n=23) relative to the percentage in the group that live with a cat (15.1%; n=8), $X(1)$=13.34, p<.01.

Multivariate Analysis

Table 5 presents a binary logistic regression analysis examining the predictors of a study participant being *Happy* (Yes/No). Data indicated that the overall model was statistically significant, $X(4)$=58.56, p<.001. Furthermore, data indicated that 85.0% of cases were categorized correctly. In terms of individual predictors, in reference to those with a high school/GED level of education, those with a graduate school level of education were almost 24 times (OR=23.85, 95% CI=4.19-135.67) more likely to be *Happy*, X^2 (1, $N = 100$) = 12.78, *p*<.001. Findings also indicated that a higher income level was also significantly associated with being *Happy*, X^2 (1, $N = 100$) = 9.00, *p*<.01. Lastly, data indicated that study participants that live with a dog are over nine times (OR=9.55, 95% CI=2.24-40.67) more likely to be *Happy*, relative to study participants that live with a cat, X^2 (1, $N = 100$) = 9.32, *p*<.01

DISCUSSION

The study findings supported the study hypotheses that postulated that study participants that live with a dog will be significantly more likely to be happy, relative to those that live with a cat. There are several possibilities for this finding...

Limitations

This current study also incorporates several limitations…

CONCLUSION

The finding that study participants that live with a dog are significantly more likely to be happy, relative to those that live with a cat, suggests that future research should examine this topic further. Specifically, future research might benefit from incorporating a qualitative piece that includes probing qualitative questions that might better identify the reasons behind this finding…

Table 1

Descriptive Analysis of Categorical Variables (*N*=100)

Variable	n	%
Education Level		
High School/GED	32	32.0
College	29	29.0
Graduate School	39	39.0
Do You Live with a Dog or Cat?		
Dog	47	47.0
Cat	53	53.0
Is the Study Participant Happy?		
Yes	31	31.0
No	69	69.0

Table 2

Descriptive Analysis of the Continuous Variables (*N*=100)

Variable	M (SD)	MIN/MAX	Potential Scores
Income Level	197, 250.00 (103,336.79)	100,000-495,000	NA
Happiness	2.73 (.76)	1.20-4.40	1.00-5.00

Table 3

Bivariate Analysis of the Income Level and Happy (*N*=100)

Variable	M (SD)	t (df)	p
Is the Study Participant Significantly Happy?		11.6 (98)	.000
Yes	$313,709.68 (101996.14)		
No	$144,927.54 (44676.72)		

Table 4

Bivariate Analysis of the Education Level, Do You Live with a Dog or Cat, and Happy (*N*=100)

| | Is the Study Participant Happy? | | | | |
| | Yes | | No | | |
Variable	n	(%)	n	(%)	X^2(df)
Education Level					28.01 (2)**
High School/GED	3	(9.4)	29	(90.6)	
College	4	(13.8)	25	(86.2)	
Graduate School	24	(61.5)	15	(38.5)	
Do You Live with a Dog or a Cat?					13.34 (1)**
Dog	23	(48.9)	24	(51.1)	
Cat	8	(15.1)	45	(84.9)	

**p<.01

Table 5

Binary Logistic Regression Analysis Examining Happy (*N*=100)

Variable	B (SE)	Wald (X²)	OR (95% CI)
Education Level			
High School/GED (Reference Group)			
College	.33 (.95)	.12	1.39 (.21-9.00)
Graduate School	3.17 (.89)	12.78	23.85 (4.19-135.67)***
Income Level	.04 (.01)	9.00	1.04 (1.01-3.17)**
Do You Live with a Dog or Cat?	2.26 (.74)	9.32	9.55 (2.24-40.67)**

Note. For Model: $X(4)$=58.56, $p<.001$. **$p<.01$, ***$p<.001$.

PART 6

ASSESSING PUBLISHED QUANTITATIVE RESEARCH STUDIES

6.1   What Will this Section Tell Us?

In this section we will discuss several of the factors that impact the reader's aptitude to effectively assess the quality of a published quantitative study. First, we will discuss the need for the reader to have a predetermined set of criteria in mind that specifies the needed components of a quantitative study prior to reading a specific work, which can be facilitated by applying *The 7 Steps of Data Analysis* model. We then examine how the mind has a way of examining findings before substantiating that the methods are sound, which is a common mishap in evaluating a published study. Next, we will illustrate how a small methodological error can produce a big problem in analysis, how you do not need to be an expert in statistics to effectively evaluate statistical research, and the utility of applying a *strengths-based* perspective in assessing published quantitative research. Lastly, we illustrate how to apply *The 7 Steps of Data Analysis* model as an effective means of assessing a published quantitative study.

6.2   Predetermined Criteria: The Key to Assessing a Quantitative Study

The aptitude to review and assess the quality of a published quantitative study is a great skill, but one that takes effort and strategy to learn. One of the strongest assets a reader can have toward effectively assessing a published data analysis study is a set of *predetermined criteria* reflecting the necessary components that must be included in an article or report. Certainly, a key to assessing any work is the ability to describe if all essential information has been included. Furthermore, when all necessary information has not been included, a reader should also be able to articulate what important elements are missing.

To achieve either of these goals, the reader must know all the necessary elements that must be included in a quantitative study prior to reading a specific work. Although, few readers approach published quantitative research with specific criteria in mind that must be satisfied before he or she deems that study credible. Typically, a reader observes a study and judges the quality of the materials presented, *but does not make a rigorous assessment reflecting if all the necessary materials were present.*

For example, as reviewed earlier in the text, the presence of *missing data values* can have a far reaching impact on study findings. However, a description of missing data values is often not presented in published research. If readers approach a quantitative study *with the conscious thought that missing data is an important issue* that must be addressed in a published study, this <u>would be a cause for concern</u>. However, if a reader does not approach published research with this thought in mind, a neglect to describe how data were or were not missing within a published quantitative study would likely be of <u>no consequence at all</u>.

This absence of these predetermined criteria reflecting the necessary components of a published study is often a source of concern for the reader. For example, most readers would agree with the premise that study findings should not be regarded as credible until the methods are judged as sound. However, if the reader does not have a list of criteria that must be met to determine if the published study is methodologically appropriate, he or she may have trouble identifying which published studies are sound, unsound, or somewhere in between. In essence, the reader is left with a sense of uncertainty regarding which study findings should be regarded as credible and to what degree.

Therefore, there is a significant need for a strategy that would inform readers of the essential components that must be presented in a published quantitative study toward assessing the quality of that work. *The 7 Steps of Data Analysis* model, may offer readers such a strategy. Specifically, this model presents quantitative research as a series of steps, with clear tasks that must be addressed within each step. Subsequently, with knowledge of these steps and tasks, a reader can observe if each has been addressed in a published study to his or her satisfaction. In other words, knowledge of

The 7 Steps of Data Analysis model would facilitate the reader approaching the published study with a preconceived knowledge of the necessary components of a data analysis study, prior to reading a particular work.

6.3   The Mind's Tendency to Make Sense of Findings

The need to approach a published quantitative study with a list of criteria to assess methodological soundness is exacerbated by one blatant inclination of human nature. Specifically, the human mind has a strong tendency to overlook the methods used to determine findings, and immediately begins to assess the relationships described between variables. At times, it seems as if the human mind is not wired to incorporate a *look before you leap* mentality in terms of assessing if findings were produced in a responsible manner, before results are considered. Therefore, to combat this tendency of the mind, there is a need to approach published research with a well thought out list of criteria for methodological soundness that must be met before study findings are considered.

Some may wonder what tendency of the human mind is being referred to in this section. Therefore, we will now provide a short illustration of how the human mind tends to make sense of any developments, behaviors, and situations presented, prior to assessing the validity of the supporting data. Indeed, there is often an almost complete absence of a filter questioning if the information presented is valid, before the mind goes to work in an attempt to make sense of the information. As an illustration, let's say one day at work you notice a coworker who is usually happy and upbeat seems unhappy. In such instances, the mind typically begins to speculate about the sources of this seeming unhappiness.

For example, the mind may suggest the coworker had a fight with his or her spouse or received a speeding ticket on the way to work. In reality, that coworker might be as happy as usual. However, the mind does not need to wait for small details, such as verifying that the perceived observation is valid or not valid. In such instances, the mind will immediately begin to make up reasons describing why that coworker is unhappy today, even if that perception is in no way based in reality.

This tendency of the mind is easily generalized to the findings of published quantitative research. For example, if a study reports that *50% of population X is depressed*, the first thought of the mind is typically *why?* After the question of *why?* is broached, a long list of reasons for the finding often follow. However, a more sophisticated response might be, *how was this value of 50% computed?* Admittedly, the second response involving taking a step back goes against human nature, but it is most appropriate.

Therefore, in this section, we will practice the art of taking a step back to assess the methods used to compute a statistical finding, before that statistical finding is even considered as an actual credible finding. Next, we will review the manner in which statistical research articles/reports are typically interpreted by readers today, as well as suggest some techniques of effectively assessing the quality of these works.

6.4   Small Errors and Big Problems

At this point, many readers might be wondering how important is it to *look before you leap* in terms of assessing the methodological soundness of a study before findings are considered. The answer of course is a resounding *very important*. Be certain, statistical findings can very easily be skewed, misleading, and misrepresentative of the data as a result of one small methodological error (Barnett and Lewis, 1994). These small errors can easily *hide a true finding when a true finding does exist*, as well as *indicate a true finding when one does not exist* (i.e., produce a *false finding*).

To illustrate the *incredible ease* with which findings can be impacted by one small methodological error, let's consider what might happen if one simple check of data integrity (**Step 3: Checks of Data Integrity**) was not observed. Specifically, let's consider how a failure to assess the assumption of **no undue influence of outlier scores** (as described in **Section 4.4.2.5**) may result in the study research question and hypothesis being answered inaccurately. To be exact, *we will examine how the results of sample study one would change with the addition of six outlier scores*. Through this analysis, we will illustrate how the influence of a small number of outlier scores can contort study results. Simultaneously, we will demonstrate how the failure to make the necessary checks for an *undue influence of outlier scores* will result in findings that misrepresent the data.

Recall within **Section 4.6.3** within sample study one, we used an independent-samples t-test to identify a true finding where among the 100 study participants the mean score reflecting *Happiness*, was higher among study participants that lived with a dog ($M=3.02$, $SD=.79$) relative to those that lived with a cat ($M=2.48$, $SD=.62$), $t(98)=3.8$, $p<.001$. We consider this a true finding because we carefully applied all the necessary checks of data integrity (e.g., tests of normality, homoscedasticity, and no significant outliers) to the data prior to analysis. In short, we scrutinized the data to be sure they were appropriate for the independent-samples t-test. Therefore, we had confidence in the findings produced.

However, consider what might have occurred if we had skipped just one of the necessary checks of data integrity, such as assessing there is **no undue influence of outlier scores** regarding the continuous dependent variable *Happiness*. For example, instead of 100 study participants, let's suppose we had an extra six cases (now a total of 106 cases) that were significant outlier scores. Specifically, suppose we had **three extra study participants that live with a cat**, all of whom evidenced the highest score reflecting *Happiness* (i.e., scores of **5.0**). Likewise, suppose the **three other extra study participants lived with a dog**, but each of them evidenced the lowest score reflecting *Happiness* (i.e., scores of **1.0**).

Suppose we added these six cases to our original normally distributed sample of 100 scores, then repeated the independent-samples t-test with this new sample of 106 study participants. Next, we again apply the independent-samples t-test to address the central research question/hypothesis examining if study participants that live with a dog have a higher mean score of *Happiness*, relative to those that live with a cat. However, rather than the original normally distributed sample of 100 scores, we now have 106 scores, six of which are outlier scores.

SPSS Output 6.1 presents the results of the independent-samples t-test, which is identical to the original independent-samples t-test presented in **Section 4.6.3**, *with the exception of the six outlier scores being added to the sample*. Specifically, in the first box within the output, you can see that *three scores* have been added to those in the group of study participants that live with a dog (from 47 to 50), as well as those that live with a cat (from 53 to 56). Also in this box, notice the mean *Happiness* scores have changed significantly from the first analysis for those who live with a dog or cat (respectively) from 3.02 (*SD*=.79) and 2.48 (*SD*=.62) to the current 2.90 (*SD*=.91) and 2.61 (*SD*=.83). In other words, the mean scores for each group have become much more similar with the addition of these six outlier scores.

Furthermore, notice within the **SPSS Output 6.1** that within the independent-samples t-test with the 106 study participants, the value in the *Sig (2-tailed)* column is .094. This value indicates that the level of probability is greater that .05 ($p>.05$), which conveys that the difference between the mean *Happiness* scores is no longer statistically significant (i.e., there is no difference in *Happiness* scores among study participants that live with a dog or cat). The addition of these six outlier scores introduced a methodological error that has hidden the true finding revealed in sample study one.

Notice, if a data analyst had conducted this independent samples t-test without checking for outlier scores (i.e., the six outlier scores we added), he or she would have come to the conclusion that the dependent variable (*Happiness*) was not significantly associated with the independent variable (living with a dog or cat). Those results would have likely been reported in the manuscript. Thus, just this one methodological error would have resulted in the presentation of an incorrect statistical finding regarding the central research question and hypothesis.

SPSS Output 6.1

T-Test

Group Statistics

	Do You Live with a Dog or Cat?	N	Mean	Standard Deviation	Std. Error Mean
Happiness	Dog	50	2.9000	.90914	.12857
	Cat	56	2.6143	.8337	.83371

Independent Samples Test

No longer statistically

		Levene's Test for Equality of Variances		t-test for Equality of Means					95% CI of the Difference	
		F	Sig	t	df	Sig. (2-tail)		SE Diff.	Low	Up
Happiness	Equal Variances assumed	1.8	.12	1.7	104	.094	.29	.17	-.05	.62
	Equal Variances not assumed			1.7	99.9	.096	.29	.17	-.05	.62

Consider, even if every other aspect of the data analysis study was conducted absolutely perfectly (e.g., study map, data entry, checks of data integrity, the univariate, bivariate, and multivariate analysis, and write-up), with the exception of this one methodological error, this error would have resulted in findings that misrepresented the study data. Specifically, this one methodological error would have resulted in the entire study being a work that will not benefit the field, but misrepresent the central relationship being examined. In research, very often it is one simple honest mistake in a sea of virtuous hard work that results in study findings being misrepresented. However, in spite of the significant role these checks of data integrity play in producing accurate findings, often little attention is paid to these procedures within the published literature (Osborne, Jason & Elaine Waters, 2002).

The Other Way Around: Looking True, But Being False. It is essential to realize that just as a small number of outlier scores (e.g., our 6 outlier scores within a total of 106 scores) can make a true finding appear false, these outliers can just as easily make a false finding appear true. Specifically, the true finding might have been that among the 100 study participants there was not a statistically significant finding between *Happiness* scores among study participants that live with a dog or cat. However, the addition of six outlier scores might have made the difference appear statistically significant. Thus, this too would have not been a true finding, but a finding that is an aberration based on errors.

Thus, it is essential to realize that many findings presented as statistically significant true relationships in published quantitative research are actually due to errors in the data, such as a small number of outlier scores that have not been identified. Furthermore, if the methods in a published quantitative study do not describe if checks of data analysis were performed, the reader cannot be certain if the findings presented are true findings or are findings based upon errors in the data. In addition to outlier scores, errors in the data that might misrepresent findings can also be based upon the other checks of data integrity, including the other test assumptions, statistical power, missing data, and measurement tools.

Lastly, consider that this error might not seem to have dire implications as we are examining the continuous variable *Happiness* among two groups, study participants that live with a dog or cat. However, we could just as easily be applying the same independent-samples t-test to examine a continuous variable with more serious implications, such as *Recovery time* among a group of patients that receive and do not receive a new intervention. If this scenario where outlier scores were not examined occurred here, the findings could have easily indicated that an intervention that was successful was not effective or that an intervention that was not successful was effective. That is why it is recommended that before findings are considered credible, a reader affirms that the study is methodologically sound.

6.5 You Don't Need To Be a Statistician to Evaluate Statistical Research

In this section we have described that knowledge of statistical procedures is a prerequisite to being able to identify the level of credibility of a published quantitative study. However, it is also important to mention that expertise in statistics is not a requisite for critically evaluating a quantitative study. In fact, even a reasonable familiarity with *The 7 Steps of Data Analysis* model can be a great benefit in evaluating the credibility of statistical research. For example, I once introduced *The 7 Steps of Data Analysis* model to a doctoral level class of nursing professionals at the beginning of a spring semester class (in January), to aid them in their evaluation of published peer-reviewed quantitative studies. I returned to the class in March to respond to any questions that might have developed as they applied the model to their assigned readings. About half way through my visit one of the students who had never taken a statistics or research class before that semester raised her hand. She then informed the class of how she had applied *The 7 Steps of Data Analysis* model in critically evaluating quantitative research.

She first informed the class that she had observed another researcher conducting a linear regression analysis using SPSS (i.e., **Step 6: Multivariate Analysis**) at her place of employment. The student then asked the researcher what she did to prepare the data prior to conducting multiple linear regression. The researcher then told her that she was unaware of how to prepare the data for analysis in that manner (although she was aware these techniques exist), but simply was given the data, conducted regression, and reported the results. Essentially, the researcher told the student that **Step 3: Checks of Data Integrity** was skipped entirely.

Although skipping **Step 3: Checks of Data Integrity** is common as we illustrated in this section, this action can easily lead to study findings being misrepresented. For example, multiple linear regression analysis assumes no outlier scores and is therefore susceptible to being impacted by outliers in the same way the independent-samples t-test. So through skipping **Step 3: Checks of Data Integrity** while conducting multiple linear regression analysis, the researcher was susceptible to the same outlier score effect produced where we added the six outlier scores to the second t-test. Thus, when that researcher conducted multiple linear regression analysis, she had no way of telling if the results being produced were true findings based on variable relationships or false findings based on errors in the data.

However, because the student had been exposed to *The 7 Steps of Data Analysis* model, she not only knew to ask about data preparation, but was aware that the statistical findings from the analysis might be questionable because an essential step of data analysis was neglected. Please note, this student did not know how to perform **Checks of Data Integrity (Step 3)**. She also may not have even been able to identify the specific procedures that needed to be conducted to check those specific data for integrity. However, she had grown capable of making one critical distinction. Through her knowledge of *The 7 Steps of Data Analysis* model, she was able to identify that an essential step of data analysis (**Step 3: Checks of Data Integrity**) had been neglected and that the findings of the associated statistical research might be less than credible.

Practically speaking, most people who learn statistical procedures and analysis through their academic coursework do not become professional data analysts and statisticians. Thereby, these learners do not utilize, practice, or reinforce their learning through regular application of the statistical procedures. In turn, many learners question the utility of learning data analysis if the skills might soon fade away because the materials are not being applied.

However, the student experience described above might help illustrate this utility. Specifically, even if the knowledge of how to perform certain statistical procedures is not retained after academic coursework is completed, knowledge of the fundamental steps involved in conducting a data analysis study likely will be. Thus, as illustrated in the student's experience, this knowledge can quickly begin to separate credible research from less than credible research.

6.6   A Strengths-Based Perspective

When reading a data analysis study, a strengths-based perspective is often instrumental in getting value from that study. The strengths-based perspective is commonly used in the practice of social work. This approach is based upon the assumption that all clients have strengths. The goal of the approach largely entails identifying, focusing upon, and developing the specific strengths of each individual client. Through focusing upon specific strengths in treatment, these qualities are ideally brought from the abstract into the physical world. Many times, it seems that client strengths might have gone unseen and unused if not for this strength focused approach. Naturally, as with any comprehensive treatment approach, the strengths-based perspective also involves listing the client characteristics that may require revision rather than accentuation.

A key in deriving value from a published data analysis study is to approach the published study with many of the same guidelines incorporated in the strengths-based approach. First, if you assume that each published data analysis study has strengths, you will be more likely to identify those strengths. For example, many times the strengths of a data analysis study are not immediately apparent, at which point the reader might prematurely dismiss the study. However, if you have the underlying assumption that all studies have strengths, you might search a bit harder, which might result in your finding significant value in the study that would have been otherwise lost.

As mentioned above, within the field of social work, the strengths-based perspective involves the identification of the specific strengths of each individual client. Data analysis studies also have highly individualized strengths that need to be considered. For example, one study might be absolutely technically correct, which would be a strength. Another study might not be entirely technically correct, but might have a strength where the work is the only study that examines a certain topic or outcome. The identification of and focus upon the individual strengths of a study enhances the contribution that study might make to your learning and research.

However, as previously mentioned, the strengths-based approach must also incorporate a careful consideration of traits that might present a challenge. There is often a fine line one must walk when reading data analysis studies where one notes study strengths, but also problematic areas that could invalidate all the positive aspects of the study. For example, a study might be incredibly innovative, but fail to describe if data were prepared properly for analysis. Thus, the premise of the study might be a significant asset, but at the same time the lack of data preparation make the findings dubious. Ideally, the strengths-based approach would result in a careful examination of the strengths of a study, while balancing those attributes with any aspects of the analysis that might be problematic.

6.7   Assessing a Quantitative Study

The published peer-reviewed quantitative study (see the Appendix) we will examine is titled *Are barriers to service and parental preference match for service related to urban child mental health service use?* We will assess this manuscript though applying *The 7 Steps of Data Analysis* model to the article.

6.7.1   Applying Step 1: Study Map

The first step in a quantitative study is to write out the study map. Recall, the study map presents the essential relationships being studied in the data analysis. Please see the **Step 1: Study Map** sections for sample study one (**Section 4.2**) and two (**Section 5.2**) as examples. When reading published statistical research we should be able to structure the same type of study map from the information provided. At times, studies may not have certain features, such as a stated hypothesis and/or covariate variables. These types of variations between studies are common, so it is important to be able to fit the information into a study map in spite of the different ways the information might be presented. However, within each study the study purpose and relationships between variables should always be identifiable.

To illustrate this, we have applied the information from the published research study into a study map form, which is presented in **Figure 6.1**. Much of the information entered into the study map is available in several parts of the published research study. However, as we stated earlier, the purpose of the study and study variables are often clearly presented in the final paragraph before the Methods section. For example, within the published manuscript the final paragraph prior to the Methods section states:

Thus, the goals of the current study are: 1) to describe barriers that families face in obtaining child mental health services and how they relate to service use; 2) to identify if matching parental preferences for specific service type relates to ongoing involvement in child mental health services; 3) to examine whether matching parent preferences for service type for their child or types of barriers is more strongly associated with the number of sessions attended; 4) to draw implications for service providers to improve service use in urban child mental health services.

From this paragraph, we can glean the essential information to fill in the study map for the published study. We can clearly identify the research question or rather research purpose, which would have been the research question if phrased as a question. Specifically, we could describe the research question or study purpose as:

Research Question (Study Purpose)

1) To describe barriers that families face in obtaining child mental health services and how they relate to service use. We could phrase this as the research question: *What are the barriers that families face in obtaining child mental health services and how they relate to service use?*

2) To identify if matching parental preferences for specific service type relates to ongoing involvement in child mental health services. To phrase this statement as a research question: *Does matching parental preferences for specific service type relate to ongoing involvement in child mental health services?*

3) To examine whether matching parent preferences for service type for their child or types of barriers is more strongly associated with the number of sessions attended. To phrase this statement as a research question: *Is matching parent preferences for service type for their child or types of barriers more strongly associated with the number of sessions attended?*

4) To draw implications for service providers to improve service use in urban child mental health services. To phrase this statement as a research question: *What are the implications for service providers to improve service use in urban child mental health services?*

From these statements, the independent and dependent variables become apparent. The analysis is explaining service attendance (the dependent variable) as a function of parental preferences for service type and barriers to service (the independent variables).

Dependent Variable: Service attendance (1 to 16)

Independent Variable: Parental preferences for service type, barriers to service

However, within that paragraph, there is no mention of covariate variables. Typically, if covariate variables are not specified in the final paragraph before the Methods section, any covariate variables will be mentioned in the Data Analysis Plan section of the manuscript. For example, in the final sentence of the data analysis plan there is a sentence which reads:

Child age, gender, and race were controlled for in the multivariate linear regression model.

Thus, the term "controlled for" suggests that these variables were covaried in the regression model. In other words, the factors were included as covariate variables.

Covariate Variables: Child age, gender, and race

Lastly, as we read the manuscript, we see that there is no formally stated study hypothesis. Specifically, this published study is more of an observational study, where the relationships between variables are being observed.

Hypothesis: No stated study hypothesis.

Figure 6.1 Study Map for the published quantitative study

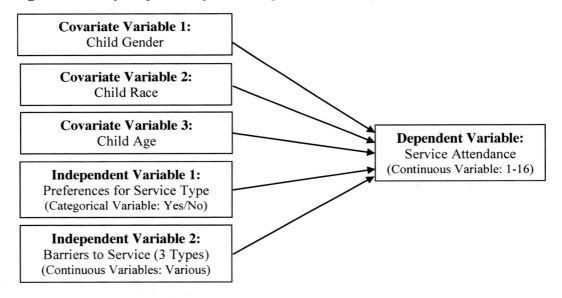

Subsequently, in terms of **Step 1: Study Map**, the published article seems to be adequate. Certainly, if we could not have filled in this information, the quality and credibility of the study might be called into question as that would suggest the research questions and/or study purpose, along with the key study variables (covariate, independent, and dependent variables) might not be apparent.

6.7.2   Applying Step 2: Data Entry

Data entry is an essential step in a data analysis project. Please see the data entry portions of study sample one (**Section 4.3**) and two (**Section 5.3**) for further information. However, although data entry is an important step in a quantitative study, the process is barely ever mentioned in a peer-reviewed journal article and many research reports, including the published manuscript examined in this section. Thus, even though the process of data entry is not described within the published research study we are examining, we will not see this as a sign of questionable quality.

6.7.3   Applying Step 3: Checks of Data Integrity

As mentioned earlier, **Step 3: Checks of Data Integrity** is a vital step in a quantitative study to the point where findings cannot be regarded as entirely credible if this step is not conducted efficiently. For further details, please see **Section 4.4** within sample study one and **Section 5.4** within sample study two. Within each of these sections we see that checks of data integrity should include examining statistical power, test assumptions, missing data, and measurement tools. In this section we will apply these recommendations and assess how well these elements were addressed in the published manuscript.

6.7.3.1   Statistical Power

Statistical power reflects our power to reject the null hypothesis when it is false, which is critically important to a quantitative study. Statistical power is often examined using a **power analysis** statistical procedure, which is generally regarded as the tool used to estimate an adequate sample size for a study (i.e., to answer the question of how many study participants you need for a study). For details on statistical power and the power analysis, see **Section 4.4.1** within sample study one and **Section 5.4.1** within sample study two. As presented in our sample manuscripts, typically a power analysis might be reported in the Data Analysis Plan or perhaps somewhere in the Results section.

However, in terms of the published manuscript, **we see there is not a power analysis presented**. Although, this lack of reporting may not have the same implications as other checks of data integrity that might not be reported. Specifically, we could conduct a power analysis to identify the number of study participants needed in a regression model required to detect an effect between variables. This estimate is not specific to the dataset, but is a general estimate computed on statistical software. However, most checks of data integrity (e.g., checks for outlier scores) are specific to the dataset and are not identifiable without the use of the actual data.

We see the published manuscript uses a multiple linear regression model with four predictors. We know from the power analysis from sample study one, with all the appropriate settings, a sample size of 85 study participants would provide sufficient statistical power to test the variable relationships. Subsequently, we see in the sample manuscript that 146 study participants were included in the multiple linear regression model. Thus, although not reported, it seems that there is sufficient statistical power to test the relationship between variables.

Remember as we pointed out in the sample studies, if the analysis does not have sufficient statistical power, a true finding might exist, but the analysis might not have enough statistical power to indicate that relationship. Thus, many times a study will indicate that a statistically significant relationship between variables does not exist, when the study actually does not have enough statistical power to indicate if that relationship does or does not exist. It is similar to a type of error in analysis. However, here the error is associated with stating that there might or might not be a true finding, when there is not sufficient statistical power to make that statement with a proper level of credibility.

For example, suppose the sample manuscript had a sample of 55 study participants. Likewise, suppose that the multiple linear regression model indicated that parental preference match for service was not related to the number of mental health sessions attended. Since we know we need 85 study participants to generate sufficient statistical power to test this relationship, we cannot identify if the association is correct because the sample size is too small to generate the statistical power to identify the relationship. Thus, such a question regarding statistical power could be associated with issues related to the credibility of findings.

6.7.3.2 Test Assumptions

As mentioned earlier, **Step 3: Checks of Data Integrity** is a vital step in a quantitative study to the point where findings cannot be regarded as wholly credible if this step is not conducted efficiently. For further details, please see **Section 4.4** within sample study one and **Section 5.4** within sample study two. We also mentioned that this step is perhaps the most commonly skipped step within the process of data analysis. The most frequent reason **Step 3: Checks of Data Integrity** is ignored is that it is entirely possible to conduct a data analysis study, write a manuscript based upon the findings, publish the paper, and have the work generally accepted by readers, without performing any of these checks.

However, as described in **Section 4.4** within sample study one, when examining a continuous dependent variable using multiple linear regression analysis, **Step 3:**

Checks of Data Integrity represents an involved and detailed process. For example, that section describes that an efficient analysis of checks of data integrity for such a study would include an assessment of a normal distribution (**Section 4.4.2.1**), multicollinearity (**Section 4.4.2.2**), homoscedasticity (**Section 4.4.2.3**), linearity (**Section 4.4.2.4**), no undue influence of outlier scores (**Section 4.4.2.5**), as well as several other test assumptions (**Section 4.4.2.6**).

However, although the published manuscript examines a continuous dependent variable via multiple linear regression (and several other parametric tests, e.g., t-test, correlation, based on similar assumptions), most of these necessary checks of data integrity are not mentioned. This is cause for concern, as we already demonstrated in **Section 6.4**, just one of these checks being skipped, such as a check of *undue influence of outlier scores* (**Section 4.4.2.5**), could result in misleading study findings. Thus, while the published manuscript does mention that checks were made for multicollinearity (in the Data Analysis plan section we see the statement: Checks for multicollinearity among independent variables revealed no significant issues) the absence of the remaining checks of data integrity is a red flag reflecting that the study results might be based upon errors in the data rather than true findings.

6.7.3.3 Missing Data

As mentioned earlier, issues surrounding missing data values in a quantitative study can have a number of implications, such as biasing study results and impacting generalizability. Please see **Section 4.4.3** of sample study one for a description of the issues and solutions regarding missing data values. As in our sample manuscripts, typically a description of missing data values will be described in the Data Analysis Plan or perhaps somewhere in the Results section. **However, we can see in our published manuscript that there is no mention of missing data values.** This is not adequate. Specifically, even if there were no missing data values within the data set, the manuscript should mention that the data were complete (i.e., there were no missing data values). The majority of studies have some missing data values that must be addressed, so an assumption of complete data might be impractical. Regarding this published manuscript, missing data values might impact findings by biasing generalizability, or in a more colloquial way of stating this issue, the question of: **Who are these findings true for?**

The published study identifies that parental match for service (the service parents wanted for their child was matched with what services the child received) was associated with greater service attendance. The finding might be true for parents under 40 years of age, but not true for parents over 40 years of age. However, suppose in data collection parents over 40 had a much higher rate of missing data values and were underrepresented in the findings. Subsequently, it might seem the finding was true for the entire population because a certain group (parents over 40) were missing from the analysis. Therefore, the finding seems generalizable to the entire population of parents, but in truth are not.

Suppose data were collected on study participant demographics for all study participants, including those that did and did not have missing data values. An analysis of missing data patterns (see **Section 4.4.3.3 Patterns of Missing Data**) might have revealed that parents over 40 years of age might have been more likely to evidence missing data values and not be represented in the statistical analysis. Therefore, the reader could at least consider that perhaps the finding might be true for parents under 40 years of age, but the findings may or may not be generalizable to parents over 40 years of age.

However, regarding the published manuscript, there is no mention of missing data values. Subsequently, we cannot be certain if there was or was not a profile type of study participants that might have been excluded from analysis due to missing data and/or if that group represented a certain profile of study participants that is underrepresented in the analysis, which might impact the generalizability of findings. In end, the anonymity regarding the missing data values could be concealing significant challenges in this quantitative analysis.

6.7.3.4 Measurement Tools

Measurement tools are of supreme importance to a quantitative study. Briefly, if study constructs are not measured adequately, the entire study will be misguided. Please see **Section 4.4.4** within sample study one for a description of issues regarding measurement tools, such as validity and reliability. Within that section we discussed that if one element of a measurement tool is mentioned, it is usually the internal consistency reliability expressed by the Cronbach's alpha. However, in the published study, there is not a scale composed of several items that is combined to measure one construct which we would use to perform the internal consistency reliability procedure.

However, it is important to note that study measurement tools composed of categorical variables (e.g., dichotomous items with the response Yes or No) that are summed to measure a construct are also subject to internal consistency reliability analysis. Specifically, an internal consistency reliability estimate (Cronbach's alpha) for a scale incorporating dichotomous items can be computed through conducting a Kuder-Richardson (KR20) reliability analysis. However, this procedure was not conducted in the published study even though the dichotomous (Yes or No) items measuring barriers were summed to reflect barrier constructs (i.e., Concrete Barriers, Stress Barriers, and Doubt Barriers). This is likely due to the fact that there are only two or three items in each of the three scales, which is a rather low number of items to use in a reliability analysis.

In terms of the published manuscript, it can be stated that scale validity and reliability is not very well assessed, but that is likely related to the instrumentation. For example, regarding composite scale reflecting the number of doubt barriers, instead of a statistical indicator (e.g., Cronbach's alpha) to reflect if the multiple items reflected a single construct is being measured, there is more of an assumption that a single construct (e.g., Doubt barriers) is being measured because the items that compose the scale seem to reflect that construct. This of course is likely related to the fact that this scale was computed using a few Yes or No items, which did not facilitate a sophisticated tool assessment. If the study had employed a more sophisticated measure, we may have expected a more sophisticated assessment of those measures. Of course, these rather basic measures do reflect a limitation that should be kept in mind.

Since an assessment of internal consistency reliability via the Cronbach's alpha was not practical, we might also consider scale validity. For example, consider **face validity**, which refers to if an instrument appears to be measuring the trait or construct it is intended to measure. As we mentioned, it appears somewhat reasonable to assume the items in the barriers scale appear to be measuring barriers as each item asks about a common issue that has been associated with interfering with service use in prior research (Burns et al., 1994).

Recall, **content validity** refers to the degree to which a measurement tool has a pool of items that adequately reflect the dimensions of (the content of) the trait or construct being measured. Here we might find a more serious measurement issue. Specifically, there are doubtlessly many more types of barriers in each of the three categories that could be measured with only two or three constructs. For example, if we observe the items used to reflect the composite scale of *Doubt barriers*, we see the items *Not sure if therapy is necessary* (Yes or No) and *Not sure therapy will work* (Yes or No). Certainly, many other dimensions of barriers related to doubt regarding service could have been included, such as *doubt* related to the *quality of the client/provider relationship*. Thus, regarding content validity, we might suggest the study measurement tool could be more comprehensive.

The other independent variable *Parental Preference for Services* also seems to display **face validity**, as parents were asked what services they wanted and intake were coded as having those preferences met if those categories of services were provided in treatment. However, upon deeper inspection, this form of categorization might not be as comprehensive as initially thought. For example, there is no measure that verifies if parental preference for service type changed immediately after intake or at any time during treatment. Thus, although preferences seem to be measured, the measurement tool used does not account for changes in preferences during treatment. Subsequently, upon deeper inspection, this measure may lack aspects of face validity.

This measure also reflects challenges related to **content validity** as it is likely that various dimensions of the services parents preferred were not captured by the measurement tool. Specifically, parental preference for service was measured through asking parents if they wanted and received individual therapy, family therapy, group therapy, and/or medication. These broad categories do not account for several preferences within categories that may or may not have been met.

For example, a parent may have had a preference for a medication service for their child with a specific drug treatment in mind. However, the child of this parent may have received medication services, but was treated with a different drug than the parent preferred. Thus, this parent would have been categorized as having his/her preference met as the parent wanted medication services and received medication services. However, the measurement tool did not facilitate a more detailed picture of parent preferences that would reflect these other dimensions.

The dependent variable number of sessions attended, required less scrutiny, as values are simply raw numbers indicating how many times the clients physically attended a session.

6.7.4   Applying Step 4: Univariate Analysis

Recall, in **Sections 4.5** and **5.5** within the sample studies, the purpose of univariate analysis is described as presenting the characteristics of each study variable individually. For the most part, the univariate analysis presented in the published study seemed comprehensive. For example, regarding *Barriers to Service,* each item within the barriers to service scales was presented at the univariate level (Table 1) before being summed into a composite scale. After these items were summed, each summed barriers scale was also presented at the univariate level (Table 2). The distribution of these independent variables do not appear to be normally distributed based upon the distribution of percentages, as the lower number of barriers seem to incorporate much higher percentages. Again, normality was not assessed in the published study, but this suggests that these independent variables may not be normally distributed.

Furthermore, the presentation included demographic characteristics (see the section *Sample and Setting*), as well as the percentage of parents that did and did not have service preference met (See paragraph one of the *Results* section). However, the univariate presentation of the dependent variable is not made. Specifically, the mean, standard deviation, and minimum/maximum values for the number of sessions attended were not presented. These values were presented for the group of parents that did and did not have service preferences met (see the paragraph *Parental Preference Match for Service and Ongoing Service Use*). Although, the comprehensive values for the *ongoing service use* variable overall were not presented.

6.7.5 Applying Step 5: Bivariate Analysis

Recall, **Sections 4.6** and **5.6** within the sample studies described a primary goal of bivariate analysis is often to determine which predictor variables are related to the dependent variable at a statistically significant level. Also, bivariate analysis is often used to assess any sample bias, in relation to several factors including the exclusion of study participants due to study design. Both of these approaches were taken in the bivariate analysis section of the published manuscript. Thus, we can assess how well bivariate analysis was conducted within the published manuscript through assessing if the correct bivariate test was applied, as well as if all the important bivariate relationships (each relationship between a predictor with the dependent variable) are presented. In addition, we might review how well bivariate analysis was employed to assess bias.

Examining the relationship between each predictor variable and the dependent variable. Perhaps our first question should examine if the correct bivariate tests were used to examine the relationships between variables. Recall, in **Sections 4.6 and 5.6,** *The Bivariate Test Key,* specifies that the proper bivariate procedure to apply when relating two variables according to the variable structure (e.g., continuous or categorical) of each variable. For example, the key describes when relating a continuous variable (e.g., the dependent variable *number of sessions attended*) and a dichotomous categorical variable (e.g., the independent variable *parental preference match for service Yes or No*) is an independent-samples t-test. Furthermore, *The Bivariate Test Key* specifies that the proper bivariate procedure to apply in combination with a continuous variable (e.g., the dependent variable *number of sessions attended*) and another continuous variable (e.g., the independent variable *barriers to service*) is a correlation analysis. Thus, we see in the published manuscript that these were the specific bivariate tests used to test these combinations of variables, so the bivariate analysis seems appropriate.

Furthermore, the published study did seem to adequately present all pertinent relationships between the dependent variable with each predictor (covariate and independent) variable. For example, within the published study in the Results section, within the paragraph titled *Barriers to Service and Child Mental Health Service Use*, a bivariate-level correlation analysis is presented that describes the relationships between the three types of barriers to service (i.e., independent variables) and the number of child mental health sessions attended (dependent variable):

Correlational analyses revealed that neither concrete, $r(144) = -.05$, $p = .57$, stressful, $r(144) = -.13$, $p = .12$, nor doubt, $r(144) = -.13$, $p = .12$, composite barrier scales were significantly correlated with service use.

A bivariate analysis was also presented that examined the relationship between the independent variable parental preference match for service and number of child mental health sessions attended. Specifically, within the published study in the Results section, the paragraph titled *Parental Preference Match for Service and Ongoing Service Use*, the bivariate level independent samples t-test analysis indicated:

A test between means revealed that caregivers whose children received the services they wanted at the time of intake attended an average of two treatment sessions more ($M = 7$ sessions; $SD = 4.4$) than those who received only a portion or none of the services they requested ($M = 5$ sessions; $SD = 3.7$), $t(144) = 2.99$, $p < .01$.

Examination of bias. As mentioned earlier in this text, bivariate analysis is often used to examine potential sample biases, as was illustrated in the published manuscript. For example, the published manuscript described that the original full sample involved in the study was composed of 253 study participants. Of these 253, 146 parents had preferences for service type and were included in the current analysis. However, there was also a sample of 107 parents from the original sample that were excluded from this analysis as they did not express a preference for service type. The bivariate analysis reflected if the excluded 107 study participants were or were not fundamentally different from the 146 study participants included in the study. For example, in the Results section in the final portion of the first paragraph the sentences read:

Chi-square analyses were conducted to determine whether, within the original sample, the subsample of parents with preferences for services ($n = 146$) differed significantly from the remainder of parents in the sample ($n = 107$) based on important family related variables: child race, child age, child gender, relationship of primary caregiver to child, and whether or not the family received public assistance. Analyses revealed that the two groups of families did not differ significantly by these factors.

Furthermore, in the Results section under the title *Barriers to Service*, the final sentences read:

Chi-square analyses were conducted to determine whether, within the original sample, the subsample of parents with preferences for services (n = 146) differed significantly from the remainder of parents in the sample (n = 107) based on the frequency of barriers (i.e., concrete, stressful, doubt) encountered by each group. No significant differences were found.

Subsequently, the bivariate analysis seems adequate within the published study. However, keep in mind, as described earlier in the text, that the correlation and independent-samples t-test used in this bivariate analysis are parametric tests that have several test assumptions (e.g., normality, no outlier scores). However, as mentioned earlier, assumptions were not examined, which may suggest the results of the tests might be in question.

6.7.6 Applying Step 6: Multivariate Analysis

Recall, **Sections 4.7** and **5.7** within the sample studies describes a primary goal of multivariate analysis is to examine the joint effect of all predictor variables on the dependent variable. Multivariate analysis often involves examining which predictors are related to the dependent variable, as well as which predictor has the greatest effect upon the dependent variable. Thus, we can assess how well multivariate analysis was implemented within the published manuscript through assessing if the correct multivariate test was applied, as well as if the dimension of statistical significance and relationship magnitude are reflected in the findings.

Examining the joint impact of all predictor variables and the dependent variable. Perhaps our first question should examine if the correct multivariate tests were used to examine the relationships between variables. Within **Section 4.7** and **5.7** *The Multivariate Test Key* specifies that the proper multivariate procedure to apply when the dependent variable is continuous is the multiple linear regression analysis. Subsequently, we see in the published study a multiple linear regression analysis (a.k.a., Ordinary Least Squares Regression Analysis) has been applied to explain the continuous dependent variable *number of sessions attended*. Thus, the selected multivariate procedure seems appropriate.

The analysis in the published manuscript specified that the goal of multivariate analysis is to examine the joint effects of independent variables upon the dependent variable, while controlling for covariate variables (child age, child race, child gender). We see through the information provided in the Methods and Results sections of the published manuscript that this was accomplished. Specifically, within the text under the heading *Barriers to Service and Parental Preference Match for Service and Service Use* we read:

Ordinary least squares regression was used to compare the relative strength of the types of barriers to service and parental preference match for service in association with the number of child mental health treatment sessions attended in services. Regression analysis revealed that the overall model of barriers and parental preferences was significant, $R^2 = .11$, $F(2, 144) = 2.37$, $p < .05$ (Table 3). However, only parental preferences match for service significantly contributed to the model, $B = 2.22$, $\beta = .27$, $p < .01$.

Thus, the results of the multivariate analysis are presented in a clear and understandable manner, which addresses the primary research questions. Namely, within the regression model, the analysis indicated that the independent variable parental preference match for service was significantly associated with number of child mental health sessions attended, while barriers to service was not.

6.7.7   Applying Step 7: Write-up and Report

Section 4.8 of sample study one, describes the elements of an effective write-up and report for a manuscript. In this section, we will mention and apply each guideline to the published quantitative study under review section by section.

Abstract. **Section 4.8.1.1** in sample study one, describes the elements that an effective abstract should reflect (as per APA style). We see the published study conforms to these requirements in terms of both logistics and content. Regarding the content, there is a concise summary of the key research points including the research topic ("The authors sought to examine how parental preference match for service and various types of barriers to service relate to involvement in urban child mental health care"), participants (parents of children receiving outpatient mental health clinic care in an urban setting), study design (single-group longitudinal design), results ("Families who received the service parents reported wanting for their child attended on average 2 treatment sessions more, whereas barriers were unrelated to service use"), and conclusions and possible implications for future research ("Considering parent preference for child mental health service may be an effective strategy in increasing service involvement in urban child mental health care.").

There was not a detailed description of the methods, including data analysis. However, depending on the specific work, a greater or lesser focus regarding each topic might be appropriate. There were also no keywords at the bottom of the abstract, but that might be stylistic based upon the specific peer-reviewed journal.

Introduction. Recall, in section **4.8.1.2 Introduction** we mentioned an introduction should present a brief but compelling statement regarding the study topic and a clear rationale for why the issue needs to be researched. A strong introduction may describe the prior theoretical and empirical research, as well as the current knowledge regarding the study topic with clear citations. In this relatively short published paper, it seems the beginning of the manuscript presents a discussion of the dependent variable, which might be considered to be part of the literature review. Again, perhaps this is done because this manuscript is styled as a short report. However, we will address the opening text in the literature review section, as these materials are the main description of the dependent variable *child mental health service use*.

Literature Review. **Section 4.8.1.3** describes the literature review needs to be focused and finite, as well as presented as three purposeful sections, and a summation paragraph just before the methods. These materials are presented below:

1ˢᵗ Section of the Literature Review: Discuss the Dependent Variable Construct. We described how the first section might present the dependent variable by definition, how the construct is measured, and discuss outcomes related to the dependent variable.

In the published study, these aspects of the dependent variable are presented in the first section of the manuscript. For example, the opening sentence describes how the dependent variable represents an essential service for many people "Urban child mental health clinics provide crucial services to inner-city children and their families."

Next, the manuscript describes the considerable proportion of the population might benefit from the dependent variable "It has been estimated that 17% to 26% of youth could benefit from some type of mental health care nationally. However, the prevalence of mental health need among inner-city youth has been found to be even greater, approximately 24% to 40%." Then the section refers to the low proportion of individuals that are using these services "In fact, the lowest rates of service usage are found among children residing in low-income urban communities."

Thus, there is a logical progression of facts that suggests an important issue exists where the dependent variable is an essential service that is underutilized, particularly in the context of the target population (youth and families residing in an urban environment). Next, the purpose of the study is introduced in the context of the dependent variable, which is to examine what predicts the use of these services "A central issue related to providing care to children presenting with mental health symptoms is understanding what influences initial and ongoing involvement in services." Thus, this Introduction/Description of the Dependent Variable builds a presentation aimed at convincing the reader that the dependent variable is an important topic that needs to be studied in the exact context of the published study.

Section 4.8.1.3 also mentions that important outcomes should be described in relation to the dependent variable. However, if you review the section in the published manuscript, there is no reference to information or citations supporting important outcomes in relation to the dependent variable (i.e., that individuals that attend mental health treatment have better outcomes). There is a chain of thought where there are references to *elevated rates of mental health difficulties* in urban environments among children, along with particularly *low rates of mental health service use*, as well as a suggestion that youth could *benefit* from mental health service use with a citation. However, there is no hard data presented that supports a relationship where children that receive a greater level of mental health services experience better outcomes (e.g., reduced levels of behavioral problems, depression, anxiety, etc.).

2nd Section of the Literature Review: Discuss the Independent Variable Construct. Next, **Section 4.8.1.3** discusses how the second part of the literature review might describe the independent variable in a similar manner to the dependent variable in terms of definition, measurement, and outcomes linked to the variable. For example, regarding the first independent variable *Barriers to Service*, the published article describes types of barriers (logistical barriers, concrete barriers, perceived barriers). Regarding the second independent variable *parental preferences for service type*, there is a short discussion as there is not a great deal of prior research on this topic.

3rd Section of the Literature Review: Discuss the Link between the Dependent and Independent Variables. **Section 4.8.1.3** also describes that the third section in the literature review should describe the link between the independent and dependent variables. Regarding the published manuscript, this would be the link between the independent variables *parental preferences for service* and *barriers to service* with the dependent variable *number of child mental health sessions attended.* The published manuscript is rather abbreviated, which resulted in much of this discussion of these contextual independent/dependent variable links being presented within the first two sections. However, this piece may be more fleshed out here in a longer manuscript.

4th In the Final Paragraph Before the Methods, Enter the Study Map Information. The final paragraph before the methods section should concisely reiterate the study purpose(s), research question(s), and/or hypothesis(es), as well as the covariate, independent, and dependent variables. We see in the published study the final paragraph before the Methods section does list these elements as this section presents:

The goals of the current study are to (a) describe barriers that families face in obtaining child mental health services and how they relate to service use; (b) identify whether matching parental preferences for specific service type relates to ongoing involvement in child mental health services; (c) examine whether matching parent preferences for service type for their child or types of barriers is more strongly associated with the number of sessions attended; and (d) draw implications for service providers to increase use of urban child mental health services.

You can see in this text taken from the published study, that the purposes of the study are listed in a lettered format. You may also note that in lines (b) and (c) it is clear that the dependent variable being explained is the number of child mental health sessions attended. Furthermore, line (b) clearly presents that matching parental preferences for services is a key independent variable, while line (c) specifies that the independent variable barriers to service will be examined as a predictor of the dependent variable relative to matching parental preferences for service.

Methods. As stated in **Section 4.8.1.4**, the Methods section should be written with a level of clarity that would allow another researcher to duplicate the study. Specifically, the Methods section should present the items listed in this section, *which are also applied to the sample peer-reviewed article:*

Recruitment of Study Participants. **Section 4.8.1.4** describes the Methods where the study participants came from (e.g., hospital setting, university setting, community setting, etc.). Accordingly, the source of the sample is described in the published study in the first paragraph under the Methods section titled Research Design:

The current study includes a sample of youth and their families from a National Institute of Mental Health (NIMH)-funded prospective study of 253 consecutively referred inner-city youth at an urban outpatient child mental health site.

Informed Consent and IRB Approval. The Methods should describe how the process of informed consent was implemented, as well as that IRB approval was secured. Both these elements are described in the published study in the second paragraph under the Methods section titled Research Design:

Consent for this study took place within a two-stage process. First, informed consent materials were read over the telephone to adult caregivers of youth consecutively referred to the clinic, and verbal consent was obtained. When families arrived at the clinic for subsequent visits, caregivers and youth signed written informed consent materials. Institutional Review Board approval was obtained.

Data Collection. The precise manner in which data were collected should also be described. The data collection is described in the published study in the first paragraph under the Methods section titled Research Design:

Data were gathered during an initial telephone contact with the adult caregivers and 4, 8, and 16 weeks after the initial contact with the clinic. Information regarding number of contacts by children and parents with the clinic was retrieved from the agency computerized scheduling system.

A description of the sample (e.g., demographic characteristics). **Section 4.8.1.4** also denotes a need to describe the sample in the Methods section, which is included in the published study in the second part under the Methods section titled Sample and Setting:

Of the families in this study, 79% (n = 201) were African American, 12% (n = 30) were Latino, 7% (n = 17) were White, and 2% (n = 5) were of other racial categories. The average age of children brought in for mental health services by their caretakers was 10 years (SD = 4.1 years). Sixty-nine percent of the children brought to treatment were male. Fifty-eight percent (n = 147) of the primary caregivers were mothers, 18% (n = 45) were grandmothers, 14% (n = 36) were foster mothers, 7% (n = 17) were aunts, and 3% (n = 8) were otherwise related. Eighty-nine percent (n = 224) of the families received public assistance. Fifty-four percent (n = 136) of the families were referred to the clinic by a community-based agency, 15% (n = 38) were referred by a friend, 11% (n = 29) by their child's school, 7% (n = 19) by a relative, 4% (n = 9) by a former patient of the clinic, and 9% (n = 22) by other sources.

Measurement instruments. **Section 4.8.1.4** also describes that a description of measurement instruments should be included in a manuscript that includes references, as well as available notations of the aspects of reliability and validity evidenced by the measurement tool in prior research. Within the published study, the measurement tools are described in the third part in the Methods section, titled Measures, which reads:

Barriers to service. Barriers to service were measured by asking parents during an intake interview whether a series of potential obstacles were anticipated to be a problem in their family's participation in child mental health services. These anticipated barriers tapped three types of barriers: concrete barriers (e.g., lack of transportation, child care, time to bring child for treatment), stressor barriers (e.g., lack of social support or exhaustion), doubt barriers (i.e., adult caregivers were not sure therapy was necessary or would be helpful). The items within each type were computed into a composite score. These items were adapted from the Child and Adolescent Services Assessment Parent Interview (Burns et al., 1994).

Parental preference for services. Parental preference for services was measured as a dichotomous variable (preference met/preferences not met). If parents received all of the services they indicated wanting for their child at intake (e.g., individual therapy, family therapy, group therapy, medication), their preferences were categorized as being met. If parents received a partial amount or none of the services for their child they indicated wanting at intake, their preferences were scored as not met.

Involvement in child mental health services. Service attendance from 1 to 16 possible sessions was retrieved from the clinic's computerized scheduling system.

Research design. The research design involved in the published manuscript was also described. For example, in the abstract, it was stated that the study involved: A single-group longitudinal design.

Data analysis plan. **Section 4.8.1.4** also describes that the data analysis plan within a manuscript should present the process applied to examine the data within a quantitative study. Ideally, all essential steps of data analysis would be described, particularly **Step 3: Checks of Data Integrity**, **Step 4: Univariate Analysis**, **Step 5: Bivariate Analysis**, and **Step 6: Multivariate Analysis**. The data analysis plan in the published study is the final paragraph in the Methods section titled Data Analysis. Many times, as in the published study, Step 3: Univariate Analysis (Descriptive Analysis) is conducted, but not formally mentioned as part of the data analysis plan. Thus, the data analysis plan often begins at Step 5: Bivariate Analysis, as is the case below. Specifically, we see in the data analysis plan for the published study that there are three phases of data analysis. The first phase and second phase involve bivariate analysis, while the third step described Step 6: Multivariate Analysis (Ordinary Least Squares, a.k.a., Multiple Linear Regression). Furthermore, there is a mention of Step 3: Checks of Data Integrity (i.e., checks for multicollinearity). However, as we pointed out earlier, the description of checks of data integrity is incomplete.

Data analysis involved three steps. First, we investigated whether rates of service use differed significantly at the bivariate level by the number and type of barriers to service that parents reported facing. Second, we investigated whether rates of service use differed significantly at the bivariate level according to whether parents' preferences for service type for their children matched with the service they actually received. Third, we entered all three types of barriers and parental preference match for service into an ordinary least squares regression analysis to examine which factors were associated with service use beyond other variables. Checks for multicollinearity among independent variables revealed no significant issues (Menard, 1995).

Results. The results should be presented in a systematic manner that follows the data analysis plan. For example, the published manuscript reports the three phases of data analysis described in the data analysis plan, as well as a univariate analysis of variables. Additionally, as recommended an appropriate *Publication Manual* style (APA) was observed in the presentation of these data.

Univariate Analysis. We see in the published article the Results section begins with an appropriate univariate analysis where the characteristics of each study variable are described independently. Specifically, the first paragraph described the results of the categorization of the parents that did and did not have preferences for service:

Of the 253 families involved in the original NIMH-funded study, 68% (*n* = 171) met the minimum criteria for inclusion (attending at least one mental health session). Of this group, 85% (*n* = 146) of caretakers expressed having a specific preference for type of service for their child, and their data were analyzed. Of the subsample of caretakers with service preference for their child, 45% (*n* = 66) received the services they requested from the clinic and 55% (*n* = 80) received services of other types.

The next paragraph is titled Barriers to Service and presents the number and percentage of study participants that reported a response of Yes or No for each item, before the barriers were combined into separate scales. In addition to the text below, these values were also presented in Table 1.

Table 1 describes the percentage of caretakers who reported anticipating each individual barrier to child mental health service use within each of the three categories (i.e., concrete, stressful, doubt). Of all barriers within each of the three categories, transportation problems (*n* = 55; 38%) and not enough time to bring child to treatment (*n* = 44; 30%), both within the concrete barriers category, were the most commonly reported anticipated barriers by caregivers.

The next composite scales reflecting Barriers to Service are presented once the individual items are combined. In addition to the text below, these values were also presented in Table 2.

Table 2 describes the barrier composite scales and frequency of barriers that caretakers anticipated facing within each of the three categories of barriers to child mental health services. Of the 146 caretakers in the current study, 55% (*n* = 80) reported at least one concrete barrier, 38% (*n* = 56) reported at least one stressful barrier, and 23% (*n* = 34) reported at least one doubt barrier that may impede their family's use of child mental health services.

Bivariate Analysis. In line with the stated data analysis plan, the bivariate analysis first presents whether rates of service use differed significantly by the number and type of barriers to service that parents reported facing:

Barriers to Service and Child Mental Health Service Use

Bivariate-level statistics were used to examine the relationship between barriers to service and the number of child mental health sessions each family attended (see Table 2 for a

summary of each of the three composite scales measuring types of barriers to service).
Correlational analyses revealed that neither concrete, $r(144) = -.05$, $p = .57$, stressful, $r(144) = -.13$, $p = .12$, nor doubt, $r(144) = -.13$, $p = .12$, composite barrier scales were significantly correlated with service use.

Also in line with the stated data analysis plan, the bivariate analysis next examined if rates of service use differed significantly according to whether parents' preferences for service type for their children matched with the service they actually received.

Parental Preference Match for Service and Ongoing Service Use

At the bivariate level, parental preference match for service was significantly associated with the number of treatment sessions attended. A test between means revealed that caregivers whose children received the services they wanted at the time of intake attended an average of two treatment sessions more ($M = 7$ sessions; $SD = 4.4$) than those who received only a portion or none of the services they requested ($M = 5$ sessions; $SD = 3.7$), $t(144) = 2.99$, $p < .01$.

Multivariate Analysis. Lastly, as stated in the data analysis plan, the final phase of data analysis involves applying an ordinary least squares (linear) regression analysis to model the dependent variable (number of sessions attended) as a function of all three types of barriers and parental preference match for service. In addition to the text below, these values were also presented in Table 3.

Barriers to Service and Parental Preference Match for Service and Service Use

Ordinary least squares regression was used to compare the relative strength of the types of barriers to service and parental preference match for service in association with the number of child mental health treatment sessions attended in services. Regression analysis revealed that the overall model of barriers and parental preferences was significant, $R^2 = .11$, $F(X, X) = 2.37$, $p < .05$ (Table 3). However, only parental preferences match for service significantly contributed to the model, $B = 2.22$, $\beta = .27$, $p < .01$.

Discussion. **Section 4.8.1.6** within sample study one describes that the author should evaluate and interpret the findings of the study in the Discussion section. The Discussion section should begin with a statement of support or nonsupport for the study hypotheses (if applicable) relative to the study findings. We see this in the published manuscript as:

Matching parental preference for service type for children receiving mental health care was significantly associated with the number of treatment sessions attended above and beyond the three types of barriers to service (i.e., concrete, stressor, doubt) that parents reported. A great deal of past research aimed at explaining child mental health service use has legitimately focused on how barriers have been correlated with reduced treatment session attendance. However, in the current study, matching parental preference of service type for their child was significantly associated with the number of treatment sessions attended. Hence, interventions aimed at increasing service use in child mental health services may benefit from considering matching parental preferences in their program design.

Section 4.8.1.6 also describes that the Discussion section should incorporate a Limitations Section where the shortcomings of the study design or some other feature of the study might be highlighted. We see this in the published manuscript as:

In the current study, parents were never asked after they began services whether their provider had explained the reasons for the change in type of service. Consequently, this study could not identify the impact of some therapists' attempts to strengthen parents' belief in the alternate services by defining the reasons for the change or whether parents who never received an explanation believed the service would not work and stopped coming for services. Thus, in future studies, more information needs to be gathered as to how well the parents were informed about how specific types of therapy might benefit their child.

Additionally, changes in parent preference for child mental health service during treatment should also be observed in this type of examination of service use. It is possible that parents' initial preference of mental health service for their child may change during the course of treatment because of factors such as changes in their knowledge of services or their child's success in treatment in accordance with certain treatment plans. The motivation of parents to continue bringing their child to services may be very different for those who experience persistent frustration as a result of not receiving the initial service they desired compared with those who may decide during treatment that a service alternate to their initial preference is more effective. Therefore, parents' posttreatment preference for child mental health services may also be an important consideration for tracking in the practice environment and in future studies of how parental preference match for service relates to service use.

Conclusion. In Section **4.8.1.7** we discuss that the conclusion should be a reasonable and accurate statement based upon implications of the findings. It should be a statement of what the findings infer, but also be careful not to go beyond those boundaries. The second paragraph in Discussion section (please see below) acts as the Conclusion statement of the published paper. Within this text is a reasonable statement of inferences that can be made based upon study findings. For example, in the text it is stated that:

The significant relationship between parental preference match for service and length of involvement in child mental health care highlights the need for examining the role of parents in the child mental health service-seeking process.

Furthermore the text states "Clearly, this issue needs to be explored further in future studies." This is a conservative recommendation, where the authors conclude that perhaps this smaller study is not conclusive, but the findings do suggest a trend that might exist and warrant further consideration.

The significant relationship between parental preference match for service and length of involvement in child mental health care highlights the need for examining the role of parents in the child mental health service-seeking process. More specifically, providing a child with a service other than that requested by their parent might be undermining the motivation of the parent or lead to their lack of belief in service efficacy and a reduced length of stay. Clearly, this issue needs to be explored further in future studies. However, on the basis of the current findings, service providers might consider providing more explicit opportunities for parents to discuss service options available in detail. For example, a short description of the various modalities of therapy could be supplied to the parent by the service provider with available effectiveness data. This might provide an opportunity for service providers and parents to agree on services needed and further enhance ongoing service involvement.

6.7.8 Summing Up the Published Article

Often the most important and complex step in assessing a published quantitative research article or report is to make a definitive conclusion regarding the overall quality, credibility, and contribution of the research. For example, even the best studies have some potential holes, flaws, or unaccounted for possibilities that could compromise the findings of the quantitative study. Likewise, even the most poorly executed studies typically have some small grain of value that can be identified. So the answer to the question of study quality, credibility, and contribution is not black and white (as in the research is either good or not good), but a matter of gradation and judgment. Thus, assessing a quantitative research study is a presentation of the classic statement *I have some good news and bad news.* Below, we will apply this approach to assessing the published research article presented above.

The Good News. Within the published article we reviewed in this section, we saw an analysis that examined how engagement in urban child mental health services is influenced by parents receiving the services they wanted for their child at intake. The first important fact to acknowledge is that at the time this manuscript was written, there were virtually no other studies that examined the topic of increasing mental health service use (child services or otherwise) through considering the preferences of users and/or their families. In fact, since this small simple article was published, it has been cited by hundreds of other studies.

As mentioned earlier, this published research study could have been more statistically rigorous. Therefore, it is rather obvious to many that this small article has not been cited by a large number of subsequent studies as an example of a quantitative study that employed sophisticated statistical procedures. Seemingly, this study was cited because the research presented an understudied topic, which was of interest to a number of other researchers. Overall, we might judge that this article does present a degree of value to the field in which it was published. Furthermore, even though our assessment of this published article was critical of some of the statistical methods because some checks of data integrity were not employed, overall the statistical procedures that were presented were sound. In end, we could suggest that the study did employ sound statistical procedures to examine an understudied topic that seemed to be of use to the field.

The Bad News. However, we would also need to acknowledge the shortcomings of the study. Specifically, we observed that several checks for data integrity were not reported. As we saw earlier in this section, even neglecting one of these procedures, such as checks for an undue influence of outlier scores, can misrepresent the data entirely. Subsequently, we can in no way be certain that the statistical results are not fundamentally wrong and misleading. Furthermore, we also noted that the study measurement instruments (the tools used to measure the study constructs) were not as exhaustive, multidimensional, and comprehensive as we might like. Thus, not only are the results of the statistical analysis somewhat dubious, but there is also some doubt regarding how well the study constructs were measured. Briefly, even though the study does have some virtues, the vices suggest the findings be considered with caution.

When the strengths and weaknesses of a quantitative study are highlighted, as we did in the published study presented here, many readers have a tendency to desire to dismiss the study findings due to the methodological shortcomings that have been underlined. *However, it is essential to realize that the great majority of published quantitative studies have methodological flaws equal to or greater than those presented in the current published study under examination.* Specifically, if you closely scrutinize the majority of published quantitative studies, sooner or later you will identify an element within that study that could easily compromise the findings. Therefore, if you are ready to dismiss a study due to the level of methodological flaws presented in this published study, you will have very few quantitative research articles to inform your research, which is rather impractical. Thus, this process of weighing each study in terms of credibility, quality, and contribution, rather than "good or bad" becomes very important.

PART 7

CONCLUSION

7.1 A Standard Model of Data Analysis

In this section, we will discuss how the research community might benefit from the availability of a *standard model of data analysis*. In this context, the term *standard* refers to *something accepted as typical*. Thus, a *standard model of data analysis* refers to a model that is accepted as the typical method applied to conduct a fundamentally sound data analysis study.

The availability of a *standard model of data analysis* has been a needed resource within the field of quantitative research for some time. For example, as we noted earlier, statistical research is quickly moving beyond the offices of statisticians and data analysts to a wider pool of professionals who are charged with learning, evaluating, and conducting data analysis studies. Currently, these groups of hard working researchers are basing their statistical methods on a combination of their professional experiences, coursework, and mentorship. It is quite realistic to state that these researchers are doing the best they can with the resources available to them.

However, as available resources vary from person to person, there is considerable variation in the level of skill and aptitude displayed by individual researchers when conducting quantitative research. This is evident in the peer-reviewed literature as among studies, many essentials of data analysis are successfully addressed and not addressed across quantitative research. In fact, even when one considers an individual researcher, one can see his or her later articles are often more technically correct than his or her earlier pieces because he or she is developing data analysis skills as articles are being published. Unfortunately, the full pool of quantitative studies with varying levels of technical correctness leads to an uneven quality of research, as well as mass confusion among readers regarding which studies are credible.

How Can a Standard Model of Data Analysis Help? A standard model of data analysis would benefit the field of quantitative research through providing researchers with a common reference point that clearly articulates the necessary steps that must be taken when conducting statistical research. This model would serve as a foundation that would convey the essentials of data analysis to bring student and novice quantitative researchers on par with more experienced researchers much more quickly. Specifically, this model would help develop early researchers through illustrating the fundamentals of quantitative research that often takes years of experience (flush with trial and error) to learn. Subsequently, the base of research literature as a whole would become more fundamentally sound as this model would serve as a universal standard that would support all works in having an adequate level of technical correctness.

For example, within the quantitative research literature, as well as the research being immediately produced, there is a huge variation between studies in the methods used to address **Step 3: Checks of Data Integrity**. This includes studies that consider this step comprehensively, as well as studies that do not consider this step at all. This reflects a very uneven quality of research. Consider our illustration in Part 6 of this text where we demonstrated that a failure to consider one check of data integrity (i.e., checks for undue influence of outlier scores) resulted in findings that misrepresented the data entirely. However, if there was a standard model of data analysis, all quantitative researchers would be instructed on the need to address **Step 3: Checks of Data Integrity**. Therefore, instead of the common scenario where researchers publish works that neglect this step until experience points out this error, the beginning researcher would know immediately that these checks of data integrity are fundamental to conducting a sound data analysis study and can employ these techniques immediately.

The absence of a standard model of data analysis reflects the fledgling nature of statistical research. Specifically, one could speculate that well developed fields have standard methods or models of operation. For example, the field of medicine has standard methods of operation that a surgeon conducting an appendectomy is expected to follow. Specifically, it would be inappropriate for a surgeon to deviate from the predetermined *model* of operation when performing an appendectomy that he or she is expected to follow as per the medical community. Essentially, prior to performing the operation, the surgeon knows there are standard methods of operation that must be followed, as well as where to find those requisites listed (e.g., AMA website). If these requisites were not specified by a standard model of operation listed in an accessible location, the globe would be fraught with surgeons performing appendectomies in many diverse ways. In essence, without this standard model, surgeons would be performing these operations with varying levels of fidelity relative to how this procedure is meant to be performed. Thus, the presence of a recognized standard model of operation is a huge asset to patients, surgeons, and all related parties.

Similarly, the specification of a standard model of data analysis would provide like benefits to those working in and served by the field of statistical research. Specifically, a standard model of data analysis would serve as a guide that specifies the agreed upon fundamental steps that a researcher must follow to maintain fidelity to statistical research. The absence of such a model is a primary cause of the currently large number of researchers performing analysis in diverse ways with varying levels of fidelity to the statistical procedures used.

Thus, in light of the potential benefits to individual researchers as well as the pool of quantitative research studies, the identification of a standard model of data analysis, should be made a priority. The result of employing such a model on a universal basis would result in a pool of quantitative researchers that are more confident and competent, as well as a richer collection of published quantitative research articles and reports.

7.2   The Next Generation of Researchers

Perhaps the richest contribution one generation of professionals can make to the next, is to devise a method to meet the most pressing challenge that earlier generation of professionals faced in their careers. For example, in academia, an almost ubiquitous challenge described by educators, students, and administrators is that statistical research is generally not regarded as accessible, understandable, and/or rewarding.

While a tremendous amount of resources are devoted toward learning, understanding, and applying statistical research procedures, the learning rarely *clicks*. That is to say, that the teaching rarely results in large groups of learners feeling confident and being competent in understanding, interpreting, and applying the fundamentals of data analysis. Perhaps the biggest contribution this generation of researchers could make to the next is to identify the method that would make the materials taught on statistical research *click*, where learners feel they truly grasp and can reasonably apply the fundamentals of data analysis, as well as find the learning rewarding.

Thus, *The 7 Steps of Data Analysis* model is presented as a tool to make data analysis more widely understood and accessible as a contribution to not only the current, but also for future generations of quantitative researchers.

Glossary

This glossary presents abbreviated definitions of the key concepts in this text. This glossary also describes many of the terms mentioned in areas of the text, such as the statistical output, that are listed but not explained. Again, one of the objectives of this text is not to get bogged down in a level of detail that obscures one's view of the process of data analysis. Therefore, many concepts that certainly warrant a fuller explanation were suppressed. Subsequently, this glossary presents not only the main concepts mentioned in this text, but also several elements that were not expanded upon fully.

Alpha (α) level: The alpha level reflects the probability of making a *Type I error* (typically set at .05).

Adjusted R^2: The adjusted R is a modified version of R statistic. Specifically, the adjusted R estimates how much variance in the dependent variable would be explained by the predictor variable(s) if the analysis was based upon the entire population from which the sample was derived.

Analysis of Covariance (ANCOVA): The ANCOVA parametric statistical procedure blends elements of an ANOVA (Analysis of Variance) and regression. Specifically, just like the ANOVA, the ANCOVA examines if mean scores of a continuous variable differ at a statistically significant level across several groups within a categorical variable. However unlike ANOVA, the ANCOVA offers the option of covarying other continuous variables (similar to the regression procedure) toward statistically controlling for the effects of those factors upon that main relationship of interest between the initially mentioned relationship between the categorical and continuous variables.

Analysis of Variance (ANOVA): The ANOVA parametric statistical procedure that examines if mean scores of a continuous variable differ at a statistically significant level across several (typically three or more) groups within a categorical variable.

Binary logistic regression: Binary logistic regression is an approach to model the relationship between a dichotomous (a continuous variable with a two response category) dependent variable with one or more predictor variables.

Bivariate analysis: Bivariate analysis involves the statistical analysis between two variables for the purpose of determining the empirical relationship between them.

β-level: The β-level reflects the probability of making a *Type II error*.

Boxplot: A boxplot is a graphical representation of observations or scores often used to display the relationship between variables. The figure lines extending vertically from the boxes (*whiskers*) indicating variability outside the upper and lower quartiles, as well as outlier scores.

Categorical variables: A variable that classifies cases in categories. Categorical variables can be nominal (cannot be rank-ordered, such as gender or race) or ordinal (can be rank-ordered, such as ratings poor, fair, good, and very good).

Chi-square test: The term chi-square test can apply to any test statistic involving a chi-square distribution. However, people commonly use the term chi-square test to refer to the Pearson's chi-square test, which is used to test the relationship between two categorical variables.

Coefficient of Determination: The coefficient of determination is the r squared statistic. In statistical tests such as the Pearson's correlation coefficient, the r squared reflects the proportion of the variance in one variable explained by a second variable.

Confidence Interval: The confidence interval consists of a range of values that with a certain degree of confidence (such as 95% in the case of a 95% confidence interval) captures this true value of a statistic (such as an Odds Ratio) within a population.

Confounding variable: A confounding variable, often called an extraneous or hidden variable, affects the variables being studied (e.g., the independent and dependent variable relationship) so that the results of analysis do not reflect the actual relationship between the variables under investigation.

Contingency table: A table presenting a cross-tabulation between at least two categorical variables.

Covariance: Covariance is a measure of how much two variables vary or change in score relative to one another.

Covariate variable: A variable included in multivariate analysis to *control* for the impact the factor has on the independent and/or dependent variable.

Cox and Snell's R^2: The version of the R coefficient of determination presented in logistic regression.

Cramer's V: An effect size estimate of the strength of the association between two categorical variables where one of the categorical variables has more than two categories.

Cronbach's alpha: The Cronbach's alpha is a coefficient of reflecting levels of internal consistency among several continuous items that measure a single construct. Subsequently, the Cronbach's alpha is commonly used as an estimate of the reliability of a study measurement tool.

Cutoff score: A clinical cutoff score (sometimes called a boundary score) refers to a score that is selected to represent the boundary between a typical and a clinically significant level on an outcome surveyed via a study measurement tool.

Degrees of Freedom: The degrees of freedom is the number of values in the final calculation of a statistic that are free to vary. The number of degrees of freedom plays a strong role in determining significance in many statistical tests.

Dependent variable: The dependent variable (also called outcome variable) is the variable in a study we are attempting to explain, most often through relating how scores vary in relation to predictor variables (e.g., such as covariate and/or independent variables).

Dichotomous variable: A categorical variable consisting of two response categories, such as Yes or No.

Dummy-coding: Dummy-coding refers to recoding a categorical variable with more than two categories into a number of separate, dichotomous variables.

Durbin–Watson test. The Durbin–Watson test is used to detect the presence of serial correlations in the residuals (prediction of errors) within regression analysis. The test is a means of detecting if the assumption of *independence of errors* has been met.

Effect size: The effect size is a measure of the strength or magnitude of a relationship, such as the degree of change in an outcome after an experimental intervention.

Eta squared: Eta squared and partial Eta squared are effect size estimates in parametric statistical procedures such as ANOVA and other general linear models. The estimate is based upon the ratio or the *model sum of squares* and the *total sum of squares*.

Exp(B): The Exp(B) is the exponentiation of the B coefficient, which is an odds ratio that reflects the unit change of a predictor in logistic regression. A value greater than 1 indicates a greater likelihood of experiencing an event, while a value below 1 indicates a greater probability of not experiencing an event.

Experimental hypothesis: The experimental hypothesis is a statement that predicts that the treatment or experimental manipulation will cause an effect.

F-ratio: The F-ratio is a test statistic that reflects the ratio of the variance between groups to the variance within groups, i.e. the ratio of the explained variance to the unexplained variance. The F ratio is used to indicate if two variances are equal.

Factor: A factor is generally another term for an independent variable or predictor. The term is common when describing experimental designs.

Frequency distribution: A *frequency distribution* is a presentation of the values within a variable, where the frequency or count of the occurrences of values within a particular group or interval is summarized.

Generalization: Generalization refers to the aptitude of a statistical model to *generalize* findings from the sample to the population from which the sample was derived.

Goodness-of-fit: The goodness-of-fit of a statistical model describes how well the model fits a set of observations. Measures of goodness-of-fit typically summarize the discrepancy between observed values and the values expected under the model in question.

Heteroscedasticity: Heteroscedasticity (the opposite of homoscedasticity) occurs when there is unequal variance of residuals at each level of the predictor variable.

Homogeneity of variance: Homogeneity of variance occurs when the variance in one variable is approximately the same at various levels of a second variable.

Homoscedasticity: Homoscedasticity (the opposite of heteroscedasticity) is an assumption in regression analysis that assumes a homogeneity of variance where the dependent variable exhibits similar amounts of variance (variance of the residuals) across the range of values for a predictor variable.

Hypothesis: A predictive statement regarding the relationship between variables.

Independence: Independence refers to the assumption that data points are separate and do not influence one another.

Independent t-test: The independent t-test (or independent-samples t-test) examines if the mean scores on the same continuous variable differ at a statistically significant level for two unrelated groups.

Independent variable: The variable of interest that explains or predicts the dependent variable.

Interval level variable: An interval level variable is a variable that is measured along a continuum with equal distances between measurement points that reflect a greater or lesser presence of a construct, *but without a meaningful true zero*.

The Kolmogorov-Smirnov test: The Kolmogorov-Smirnov test examines if a distribution of scores is different from a normal distribution at a statistically significant level. The test is better suited for large samples.

Kruskal–Wallis test: The Kruskal–Wallis is a non-parametric test that examines if more than two independent groups differ significantly by scores upon a continuous variable. This test is the non-parametric version of the One Way ANOVA.

Kurtosis: Kurtosis is a measure of the how scores cluster in the tails of a distribution. When the tails of the distribution have a high number of scores, making the distribution rather flat, the kurtosis is referred to as a platykuric distribution. When there are a small number of scores in the tails, the distribution is referred to as lepokurtic.

Lepokurtic: A distribution related to kurtosis characterized by a small number of scores in the tails of a distribution.

Linear regression: Linear regression is an approach to model the relationship between a continuous dependent variable with one or more predictor variables, where the relationship is based upon a straight regression line.

Mann–Whitney *U* test: The Mann-Whitney *U* test is a non-parametric test that examines if mean scores differ for two independent groups. The test is the non-parametric equivalent of the independent samples t-test.

Matrix: A series of values expressed in a combination of columns and rows.

Mean: The mean is the *arithmetic mean* where the values of a number of items are summed, then that summed number is divided by the number of items summed.

Mean squares: The mean square is a measure of average variability.

Median: The median is the middlemost numerical value separating the higher half of a distribution of scores from the lower half.

Mode: The mode is the value that appears most often in distribution of scores.

Model of sum of squares: The sum of squares indicates the total amount of variability explained by a statistical model.

Multicollinearity: Multicollinearity is a condition that occurs when two or more of the predictor variables in a regression equation are correlated.

Multiple linear regression: Multiple linear regression (an extension of simple linear regression) is used to predict the value of a continuous dependent variable based on the value of two or more predictor variables in a linear combination.

Multivariate analysis: Multivariate analysis explores the association between a dependent variable and one or more predictor variables.

Nominal level variable: A nominal level variable is a variable that has two or more categories, *but cannot be ranked into an intrinsic order*.

Non-parametric tests: Non-parametric tests are a group of statistical tests that do not rely on as many assumptions as parametric tests and are often used when the assumptions of parametric data are not met, such as a normal distribution.

Normal distribution: A normal distribution is a probability distribution of values of a random variable that produces a symmetrical bell-shaped graph.

Null hypothesis: The null hypothesis is the default position relative to the alternative hypothesis (research hypothesis) claiming no relationship between variables.

Odds Ratio: An odds ratio is a measure of association between an exposure and an outcome. The odds ratio represents the odds that an outcome will occur given a particular event or exposure, relative to the odds of the outcome occurring in the absence of that event or exposure.

One-tailed test: A one-tailed test is a test option used when a directional hypothesis is considered.

Ordinal level variable. An ordinal level variable is a variable that has two or more categories that can be ranked into a meaningful order.

Outcome variable: An outcome variable, also called the dependent variable, is the variable being examined and/or explained within a study by one or more predictor variables.

Outlier score: An outlier score is a value that is very different (extremely low or high) from the other scores in a distribution.

Parametric tests: Parametric tests are a category of statistical procedures that make assumptions about the parameters (defining properties) of the population distribution(s) from which one's data are drawn, such as a normal distribution of values.

Pearson's correlation coefficient: The Pearson's correlation coefficient is a measure of the linear correlation between two variables reflecting values between +1 and −1 that indicate a positive correlation, no correlation, and/or a negative correlation.

Platykuric: A distribution related to kurtosis characterized by a large number of scores in the tails of a distribution.

Population: A population refers to the total set of individuals, groups, objects, or events to whom the research study is attempting to generalize findings.

Post hoc tests: Post hoc (Latin meaning "*after this*" suggesting that an analysis was conceived of once data collection was completed) typically involves a comparison of group means for all combination of groups. For example, a post hoc analysis in a One-way ANOVA involves a comparison of mean scores for all groups in analysis (*after* the initial ANOVA model is presented).

Predictor variable: The predictor variable (e.g., a covariate and/or independent variable) is used to predict or explain a dependent variable (also called an outcome variable).

Ratio level variable: A ratio level variable is variable measured along a continuum with equal distances between measurement points that reflect a greater or lesser presence of a construct that has a meaningful true zero.

Reliability: Reliability refers to the aptitude of a measurement tool to produce consistent findings when administered to the same entities under the same conditions.

Residual: Residual (or error) represents unexplained (or residual) variation after conducting a regression model. The residual is the difference between the observed value of the variable and the value indicated by the regression model.

Sample: The sample is the subsample of the population analyzed to make inferences about the population.

Scatterplot: A scatterplot is a summary of a set of data presenting a visual picture of the relationship between the two variables.

Shapiro-Wilk test: The Shapiro-Wilk test examines if a distribution of scores is different from a normal distribution at a statistically significant level. The Shapiro-Wilk is more appropriate for smaller samples relative to the Kolmogorov-Smirnov test.

Skew: Skew refers to the measure of symmetry in a distribution of scores.

Spearman's correlation coefficient: The Spearman's correlation coefficient is used to test to examine the relationship between two continuous variables. The test is the non-parametric version of the Pearson's correlation coefficient.

Standard deviation: The standard deviation reflects the average spread of the scores in a distribution.

Standard error: The standard error is the standard deviation of the sampling distribution of a statistic. The standard error reflects the level of variability for a statistic across samples for a population.

Standardized regression coefficient (a.k.a., Standardized beta) (β): The standardized coefficients (beta) indicates the predictor variables in a linear (multiple) regression model with the greater effect on the dependent variable.

Two-tailed test: A two-tailed test is a test option used when a non-directional hypothesis is considered.

Type I error: Type I error involves a rejection of a true null hypothesis. Specifically, type I error leads one to conclude that a significant relationship between variables exists, when in reality it does not.

Type II error: Type II error is the failure to reject a false null hypothesis. Type II error would occur when analysis indicates a significant relationship does not exists, when it actually does.

Unstandardized regression coefficient (a.k.a., unstandardized beta) (*b*): The standardized coefficients (beta) indicates the degree of score change in the dependent variable according to the degree of change in a predictor variable within the context of a linear (multiple) regression model.

Variance: Variance is an estimate of the variability (spread) of scores in a distribution.

Variance inflation factor (VIF): The VIF is an estimate of the degree of multicollinearity between predictor variables.

Appendix

Are Barriers to Service and Parental Preference Match for Service Related to Urban Child Mental Health Service Use?

William M. Bannon, Jr., & Mary M. McKay

ABSTRACT
The authors sought to examine how parental preference match for service and various types of barriers to service relate to involvement in urban child mental health care. A single-group longitudinal design was used to examine whether service use at an outpatient child mental health clinic was related to parents receiving the type of service they reported wanting for their child at intake and various types of barriers to service. Families who received the service parents reported wanting for their child attended on average 2 treatment sessions more, whereas barriers were unrelated to service use. Considering parent preference for child mental health service may be an effective strategy in increasing service involvement in urban child mental health care.

Urban child mental health clinics provide crucial services to inner-city children and their families. There is a considerable range of stressors associated with living in urban communities; these stressors disproportionately impact children of color and place them at greater risk for mental health difficulties. For example, living within low-income inner-city communities has been associated with elevated rates of delinquency (Barone, Weissberg, Kasprow, & Voyce, 1995; Tolan & Henry, 1996), aggression (Attar, Guerra, & Tolan, 1994; Brown, Horn, Heiser, & Odom, 1996; Guerra, Huesmann, Tolan, Van Acker, & Eron, 1995), and other child mental health difficulties (Black & Krishnakumar, 1998; Dryfoos, 1990; Jessor, 1993; Richman, Stevenson, & Graham, 1982; Wilson, 1987).

It has been estimated that 17% to 26% of youth could benefit from some type of mental health care nationally (Brandenburg, Friedman, & Silver, 1987; McCabe et al., 1999; Tuma, 1989). The prevalence of mental health need among inner-city youth has been found to be even greater, approximately 24% to 40% (Tolan & Henry, 1996, 2000). Prior research has indicated that, although low income minority children are more likely to demonstrate increased mental health difficulties, they are less likely to use child mental health services (Kazdin, 1993; Padgett, Patrick, Burns, Schlesinger, & Cohen, 1993; Regier et al., 1993). In fact, the lowest rates of service usage are found among children residing in low-income urban communities (Griffin, Cicchetti, & Leaf, 1993).

A central issue related to providing care to children presenting with mental health symptoms is understanding what influences initial and ongoing involvement in services. It has been estimated that 50% to 75% of the children who need mental health services either never have contact with providers or terminate treatment prematurely (Kazdin, 1993). The need to examine which aspects

1

of the service delivery process contribute to successfully involving urban youth and their families in services has been identified as a gap in children's mental health research (Jensen, Hoagwood, & Petti, 1996).

Barriers to Service in Urban Child Mental Health Services

Parents play a key role in determining the number of mental health care program treatment sessions their children attend. At the very least, parents are in most cases responsible for physically bringing their child to the sessions. Keeping a child's mental health appointment at an outpatient clinic has been found to be impeded by a series of logistical barriers, also referred to as concrete barriers, that may affect a family's participation in child mental health care. Concrete barriers such as lack of transportation or child care have been found to be significant obstacles to obtaining needed mental health care for children (Hahn, 1995; McKay, McCadam, & Gonzales, 1996; Spoth, Redmond, Hockaday, & Shin, 1996).

Perceived barriers have also been found to impact a parent's ability or decision to obtain mental health care services for their children. These include lack of social support, lack of time to attend appointments, and negative expectancies of treatment outcomes. Although studies examining the impact of perceived barriers are rare, Kazdin et al. (Kazdin, Holland, & Crowley, 1997; Kazdin & Mazurick, 1994; Kazdin & Wassell, 1999, 2000) have examined the impact of perceived barriers by families on treatment participation in outpatient care. Nock and Kazdin (2001) found that parent expectancies of outcomes in child mental health services predicted subsequent barriers to treatment participation and premature termination

from therapy. Overall, the implication is that continuity of care may be seriously compromised if the perceptions of barriers by families are high.

Parental Preferences

Consideration of client preferences in family mental health services has been associated with increases in service involvement. For example, Coatsworth, Santisteban, McBride, and Szapocznik (2001) found that brief strategic family therapy, which involves a purposefully high degree of contracting (consideration of preferences) with the family, was associated with a 71% client retention rate beyond the initial service visit compared with a control group rate of 42%. However, only a limited number of studies have been conducted that investigate service outcomes in relation to parental preferences for services in child mental health. McKay, Stoewe, McCadam, and Gonzales (1998) tested an intervention that helped to increase shared perceptions of child mental health treatment goals between parents and providers during the initial interview. This interview was associated with significantly higher retention rates ($M = 7.3$ sessions) in comparison to a control group ($M = 5$ sessions), $F = 6.36$, $p < .05$. However, studies examining how directly matching parental preferences to actual services received relates to length of stay in child mental health services have not been conducted to date.

The goals of the current study are to (a) describe barriers that families face in obtaining child mental health services and how they relate to service use; (b) identify whether matching parental preferences for specific service type relates to ongoing involvement in child mental health services; (c) examine whether matching parent preferences for service type for their child or

Volume 86, No. 1

or types of barriers is more strongly associated with the number of sessions attended; and (d) draw implications for service providers to increase use of urban child mental health services.

Method

Research Design

The current study includes a sample of youth and their families from a National Institute of Mental Health (NIMH)-funded prospective study of 253 consecutively referred inner-city youth at an urban outpatient child mental health site. Data were gathered during an initial telephone contact with the adult caregivers and 4, 8, and 16 weeks after the initial contact with the clinic. Information regarding number of contacts by children and parents with the clinic was retrieved from the agency computerized scheduling system.

Consent for this study took place within a two-stage process. First, informed consent materials were read over the telephone to adult caregivers of youth consecutively referred to the clinic, and verbal consent was obtained. When families arrived at the clinic for subsequent visits, caregivers and youth signed written informed consent materials. Institutional Review Board approval was obtained.

Sample and Setting

Of the families in this study, 79% ($n = 201$) were African American, 12% ($n = 30$) were Latino, 7% ($n = 17$) were White, and 2% ($n = 5$) were of other racial categories. The average age of children brought in for mental health services by their caretakers was 10 years ($SD = 4.1$ years). Sixty-nine percent of the children brought to treatment were male. Fifty-eight percent ($n = 147$) of the primary caregivers were mothers, 18% ($n = 45$) were grandmothers

14% ($n = 36$) were foster mothers, 7% ($n = 17$) were aunts, and 3% ($n = 8$) were otherwise related. Eighty-nine percent ($n = 224$) of the families received public assistance. Fifty-four percent ($n = 136$) of the families were referred to the clinic by a community-based agency, 15% ($n = 38$) were referred by a friend, 11% ($n = 29$) by their child's school, 7% ($n = 19$) by a relative, 4% ($n = 9$) by a former patient of the clinic, and 9% ($n = 22$) by other sources.

Measures

Barriers to service. Barriers to service were measured by asking parents during an intake interview whether a series of potential obstacles were anticipated to be a problem in their family's participation in child mental health services.

These anticipated barriers tapped three types of barriers: concrete barriers (e.g., lack of transportation, child care, time to bring child for treatment), stressor barriers (e.g., lack of social support or exhaustion), doubt barriers (i.e., adult caregivers were not sure therapy was necessary or would be helpful). The items within each type were computed into a composite score. These items were adapted from the Child and Adolescent Services Assessment Parent Interview (Burns et al., 1994).

Parental preference for services. Parental preference for services was measured as a dichotomous variable (preference met/preferences not met). If parents received all of the services they indicated wanting for their child at intake (e.g., individual therapy, family therapy, group therapy, medication), their preferences were categorized as being met. If parents received a partial amount or none of the services for their child they indicated wanting at intake, their preferences were scored as not met.

Involvement in child mental health services. Service attendance from 1 to 16 possible sessions was retrieved from the clinic's computerized scheduling system.

Data Analysis

Data analysis involved three steps. First, we investigated whether rates of service use differed significantly at the bivariate level by the number and type of barriers to service that parents reported facing. Second, we investigated whether rates of service use differed significantly at the bivariate level according to whether parents' preferences for service type for their children matched with the service they actually received. Third, we entered all three types of barriers and parental preference match for service into an ordinary least squares regression analysis to examine which factors were associated with service use beyond other variables. Checks for multicollinearity among independent variables revealed no significant issues (Menard, 1995).We entered parental preference match for service last to test our hypothesis that this variable would be more strongly associated with service use beyond barriers to service. Child age, gender, and race were controlled for in the multivariate linear regression model.

Results

Of the 253 families involved in the original NIMH-funded study, 68% ($n = 171$) met the minimum criteria for inclusion (attending at least one mental health session). Of this group, 85% ($n = 146$) of caretakers expressed having a specific preference for type of service for their child, and their data were analyzed.Of the subsample of caretakers with service preference for their child, 45% ($n = 66$) received the services they requested from the clinic and 55%

($n = 80$) received services of other types. Chi-square analyses were conducted to determine whether, within the original sample, the subsample of parents with preferences for services ($n = 146$) differed significantly from the remainder of parents in the sample ($n = 107$) based on important family related variables: child race, child age, child gender, relationship of primary caregiver to child, and whether or not the family received public assistance. Analyses revealed that the two groups of families did not differ significantly by these factors.

Barriers to Service

Table 1 describes the percentage of caretakers who reported anticipating each individual barrier to child mental health service use within each of the three categories (i.e., concrete, stressful, doubt). Of all barriers within each of the three categories, transportation problems ($n = 55$; 38%) and not enough time to bring child to treatment ($n = 44$; 30%), both within the concrete barriers category, were the most commonly reported anticipated barriers by caregivers.

TABLE 1. Description of the Types of Barriers ($n = 146$)

VARIABLE	YES (%)	NO (%)
Concrete barriers		
Transportation problems	55 (38)	91 (62)
Child care problems	25 (17)	121 (83)
Not enough time	44 (30)	102 (70)
Stressful barriers		
Family/friends give hard time about coming	12 (8)	134 (92)
Feel Too tired to come	29 (20)	117 (80)
Doubt barriers		
Not sure is necessary	14 (10)	132 (90)
Not sure therapy will work	28 (19)	118 (81)

Volume 86, No. 1

Table 2 describes the barrier composite scales and frequency of barriers that caretakers anticipated facing within each of the three categories of barriers to child mental health services. Of the 146 caretakers in the current study, 55% (*n* = 80) reported at least one concrete barrier, 38% (*n* = 56) reported at least one stressful barrier, and 23% (*n* = 34) reported at least one doubt barrier that may impede their family's use of child mental health services. Chi-square analyses were conducted to determine whether, within the original sample, the subsample of parents with preferences for services (*n* = 146) differed significantly from the remainder of parents in the sample (*n* = 107) based on the frequency of barriers (i.e., concrete, stressful, doubt) encountered by each group. No significant differences were found.

Barriers to Service and Child Mental Health Service Use

Bivariate-level statistics were used to examine the relationship between barriers to service and the number of child mental health sessions each family attended (see Table 2 for a summary of each of the three composite scales measuring types of barriers to service). Correlational analyses revealed that neither concrete, $r(144) = -.05$, $p = .57$, stressful, $r(144) = -.13$, $p = .12$, nor doubt, $r(144) = -.13$, $p = .12$, composite barrier scales were significantly correlated with service use.

Parental Preference Match for Service and Ongoing Service Use

At the bivariate level, parental preference match for service was significantly associated with the number of treatment sessions attended. A test between means revealed that caregivers whose children

TABLE 2. *Description of Composite Measures and Frequencies of the Types of Barriers to Child Mental Health Services that Families Reported Anticipating (n = 146)*

VARIABLE	*n*	%
Concrete barriers		
0	66	45
1	44	30
2	28	19
3	8	6
Stressful barriers		
0	90	62
1	38	26
2	18	12
Doubt barriers		
0	112	77
1	26	18
2	8	5

TABLE 3. *Ordinary Least Squares Regression Model Examining the Association of the Number of Child Mental Health Sessions Attended With Barriers to Services and Parental Preference Match for Service (n = 146)*

VARIABLE	*B*	*SE*	β
Concrete barriers	.16	.44	.04
Stressful barriers	−.83	.55	−.14
Doubt barriers	−.53	.64	−.07
Preference service match	2.22	.69	.27*

Note. For the model, $R2 = .11$, Adjusted $R2 = .06$, $F(2, 144) = 2.37$, $p < .05$. *$p < .01$.

received the services they wanted at the time of intake attended an average of two treatment sessions more (*M* = 7 sessions; *SD* = 4.4) than those who received only a portion or none of the services they requested (*M* = 5 sessions; *SD* = 3.7), $t(144) = 2.99$, $p < .01$.

Barriers to Service and Parental Preference Match for Service and Service Use

Ordinary least squares regression was used to compare the relative strength of the types

5

of barriers to service and parental preference match for service in association with the number of child mental health treatment sessions attended in services. Regression analysis revealed that the overall model of barriers and parental preferences was significant, $R2 = .11$, $F(2, 144) = 2.37$, $p < .05$ (Table 3). However, only parental preferences match for service significantly contributed to the model, $B = 2.22$, $\beta = .27$, $p < .01$.

Discussion

Matching parental preference for service type for children receiving mental health care was significantly associated with the number of treatment sessions attended above and beyond the three types of barriers to service (i.e., concrete, stressor, doubt) that parents reported. A great deal of past research aimed at explaining child mental health service use has legitimately focused on how barriers have been correlated with reduced treatment session attendance. However, in the current study, matching parental preference of service type for their child was significantly associated with the number of treatment sessions attended. Hence, interventions aimed at increasing service use in child mental health services may benefit from considering matching parental preferences in their program design.

The significant relationship between parental preference match for service and length of involvement in child mental health care highlights the need for examining the role of parents in the child mental health service-seeking process.
More specifically, providing a child with a service other than that requested by their parent might be undermining the motivation of the parent or lead to their lack of belief in service efficacy and a reduced length of stay. Clearly, this issue needs to be explored further in future studies. However, on the basis of the current findings, service providers might consider providing more explicit opportunities for parents to discuss service options available in detail. For example, a short description of the various modalities of therapy could be supplied to the parent by the service provider with available effectiveness data. This might provide an opportunity for service providers and parents to agree on services needed and further enhance ongoing service involvement.

In the current study, parents were never asked after they began services whether their provider had explained the reasons for the change in type of service. Consequently, this study could not identify the impact of some therapists' attempts to strengthen parents' belief in the alternate services by defining the reasons for the change or whether parents who never received an explanation believed the service would not work and stopped coming for services. Thus, in future studies, more information needs to be gathered as to how well the parents were informed about how specific types of therapy might benefit their child.

Additionally, changes in parent preference for child mental health service during treatment should also be observed in this type of examination of service use. It is possible that parents' initial preference of mental health service for their child may change during the course of treatment because of factors such as changes in their knowledge of services or their child's success in treatment in accordance with certain treatment plans. The motivation of parents to continue bringing their child to services may be very different for those who experience persistent frustration as a result of not receiving the initial service they

Volume 86, No. 1

desired compared with those who may decide during treatment that a service alternate to their initial preference is more effective. Therefore, parents' posttreatment preference for child mental health services may also be an important consideration for tracking in the practice environment and in future studies of how parental preference match for service relates to service use.

References

Attar, B. K., Guerra, N. G., & Tolan, P. H. (1994). Neighborhood disadvantage, stressful life events, and adjustment in urban elementary-school children. *Journal of Clinical Child Psychology, 23*,391–400.

Barone, C.,Weissberg, R. P., Kasprow,W., & Voyce, C. K. (1995). Involvement in multiple problem behaviors of young urban adolescents. *Journal of Primary Prevention, 15*, 261–283.

Black, M. M., & Krishnakumar, A. (1998). Children in low-income, urban settings: Interventions to promote mental health and well-being. *American Psychologist, 53*, 635–646.

Brandenburg, N. A., Friedman, R. M., & Silver, S. E. (1987). The epidemiology of childhood psychiatric disorders: Prevalence findings from recent studies. *Journal of the academy of child and adolescent psychiatry, 29*, 76–83.

Brown,W. H., Horn, E. M., Heiser, J., G., & Odom, S. L. (1996). Project BLEND: An inclusive model of early intervention services. *Journal of Early Intervention, 20*, 364–375.

Burns, B. J., Angold, A., Magruder-Habib, K., et al. (1994). *The child and adolescent service assessment. Developmental Epidemiology Program*. Durham, NC: Duke University Medical Center, Department of Psychiatry.

Coatsworth, J. D., Santisteban, D. A.,McBride, C. K., & Szapocznik, J. (2001). Brief strategic family therapy versus community control: Engagement, retention, and an exploration of the moderating role of adolescent symptom severity. *Family Process, 40*, 313–332.

Dryfoos, J. G. (1990). *Adolescents at risk: Prevalence and prevention*. New York: Oxford University Press.

Griffin, J. A., Cicchetti, D., & Leaf, P. J. (1993). Characteristics of youth identified from a psychiatric case register as first-time users of services. *Hospital and Community Psychiatry, 44*, 62–65.

Kazdin, A. (1993). Premature termination from treatment among children referred for antisocial behavior. *Journal of Clinical Child Psychology, 31*, 415–425.

Kazdin, A. E., Holland, L., & Crowley, M. (1997). Family experiences of barriers to treatment and premature termination from child therapy. *Journal of Consulting and Clinical Psychology, 65*, 453–463.

Kazdin, A. E., & Mazurick, J. L. (1994). Dropping out of child psychotherapy: Distinguishing early and late dropouts over the course of treatment. *Journal of Consulting and Clinical Psychology, 62*, 1069–1074.

Kazdin, A. E., & Wassell, G. (1999). Barriers to treatment participation and therapeutic change among children referred for conduct disorder. *Journal of Clinical Child Psychology, 28*, 160–172.

Kazdin, A. E., & Wassell, G. (2000). Predictors of barriers to treatment and therapeutic change in outpatient therapy for antisocial children and their families. *Mental Health Services Research, 2*, 27–40.

McCabe, K., Yeh, M., Hough, R. L., Landsverk, J., Hurlburt, M. S., Culver, S. W. & Reynolds, B. (1999). Racial/ethnic representation across five public sectors of care for youth. *Journal of Emotional and Behavioral Disorders, 7*, 72–82.

McKay, M. M.,McCadam, K., & Gonzales, J. J. (1996). Addressing the barriers to mental health services for inner city children and their caretakers. *Community Mental Health Journal, 32*, 353–361.

McKay, M., Stoewe, J.,McCadam, K., & Gonzales, J. (1998). Increasing access to child mental health services for urban children and their care givers. *Health and Social Work, 23*, 9–15.

Menard, S. (1995). *Applied logistic regression analysis*. Thousands Oaks, CA: Sage.

Nock,M. K., & Kazdin, A. E. (2001). Parent expectancies for child therapy: Assessment and relation to participation in treatment. *Journal of Child and Family Studies, 10*, 155–180.

Padgett, D. K., Patrick, C., Burns, B. J., Schlesinger, H. J., & Cohen, J. (1993). The effect of insurance benefit changes on use of child and adolescent outpatient mental health services. *Medical Care, 31*, 96–110.

7

Richman, N., Stevenson, J., & Graham, P. J. (1982). *Pre-school to school: A behavioural study*. London: Academic Press.

Spoth, R., Redmond, C., Hockaday, C., & Shin, C. Y. (1996). Barriers to participation in family skills preventive interventions and their evaluations: A replication and extension. *Family Relations: Journal of Applied Family & Child Studies, 45*, 247–254.

Tolan, P. H., & Henry, D. (1996). Patterns of psychopathology among urban poor children: Co-morbidity and aggression effects. *Journal of Consulting and Clinical Psychology, 64*, 1094–1099.

Tolan, P. H., & Henry, D. (2000). *Patterns of psychopathology among urban poor children: Community, age, ethnicity and gender effects.* Manuscript submitted for publication.

Tuma, J. M. (1989). Mental health services for children: The state of the art. *American Psychologist, 44*, 188–199.

Wilson, M. (1987). The Black extended family: An analytic consideration. *Developmental Psychology, 22*, 246–258.

William M. Bannon, Jr.,MSW, is doctoral student at Columbia University. **Mary M.McKay, PhD,** is professor, Mount Sinai School of Medicine. Correspondence regarding the article may be addressed to the first author at wb2005@columbia.edu.

Authors' note. This work was supported by National Institute of Mental Health Grants R03 M#55327 and T32 M#14623.
Manuscript received: June 13, 2003
Revised: January 23, 2004
Accepted: January 24, 2004

References

Allison, P. D. (1999). Multiple Regression: A Primer (Undergraduate Research Methods & Statistics in the Social Sciences. Thousand Oaks, CA: Pine Forge Press.

Barnett, V. & Lewis, T. (1994). *Outliers in Statistical Data, 3rd ed*, John Wiley, Great Britain.

Berry, W. D. (Ed.). (1993). *Understanding Regression Assumptions*. SAGE Publications, Inc.

Berry, W. D. & Feldman, S. (1985). *Multiple Regression in Practice*. Sage University Paper Series on Quantitative Applications in the Social Sciences, series no. 07-050. Newbury Park, CA: Sage.

Bland, J. & Altman, D. (1997). Statistics notes: Cronbach's alpha. *British Medical Journal, 314*, 275.

Bland, J. M. & Altman, D. G. (a) (1996). The use of transformation when comparing two means. *British Medical Journal, 312*, 1153.

Bland, J. M. & Altman D. G. (b) (1996). Transforming data. *British Medical Journal, 312*, 770.

Bland, J. M. & Altman D. G. (c) Transformations, means and confidence intervals. *British Medical Journal, 312*, 1079.

Bowerman, B. L. & O'Connell, R. T. (1990) *Linear statistical models: an applied approach*, 2nd ed. Belmont CA: Duxbury.

Cochran, W. G. (1952). The [chi-squared] test of goodness of fit. *Annals of Mathematical Statistics*, 25, 315–345.

Cochran, W. G. (1954). Some methods for strengthening the common [chi-squared] tests. *Biometrics*, 10, 417–451.

Cohen, J. (1988). *Statistical power analysis for the behavioral sciences* (2nd edition). New York: Academic Press. Cohen, J. (1992). A power primer. *Psychological Bulletin, 112*(1), 155–159.

DeVellis, R. (2003). *Scale development: theory and application.* Thousand Okas, CA: Sage.

Field, A. (2005). *Discovering Statistics Using SPSS* (2nd edition). (pp. 78-87). Thousand Oaks, CA: SAGE Publications, Inc.

Jacobson, N. S. & Truax, P. (1991). *Journal of Consulting and Clinical Psychology, 59*(1), 12-19.

Kolmogorov, A. (1933). Sulla determinazione empirica di una legge di distribuzione. *G. Ist. Ital. Attuari, 4,* 83.

Kraemer, H. C. & Thiemann S. (1987). *How many subjects? Statistical power analysis in research.* Newbury Park, CA: Sage.

Menard, S. (1995). *Applied logistic regression analysis Sage University paper series on quantitative applications in the social sciences* (pp. 07-106). Thousand Oaks, CA: Sage.

Miles, J. & Shevlin, M. (2001). *Applying Regression and Correlation: A Guide for Students and Researchers.* (pp. 74-75). Thousand Oaks, CA: SAGE Publications, Inc.

Myers, R. (1990). *Classical and modern regression with applications.* Boston: Kent.

Nicol, A. (2010). Normal distribution. In N. Salkind (Ed.), *Encyclopedia of research design.* (pp. 927-932). Thousand Oaks, CA: SAGE Publications, Inc.

Nunnally, J. & Bernstein, L. (1994). *Psychometric theory.* New York: McGraw-Hill Higher.

Osborne, J. & Waters E. (2002). Four assumptions of multiple regression that researchers should always test. *Practical Assessment, Research & Evaluation,* 8(2).

Pedhazur, E. J. (1997). *Multiple Regression in Behavioral Research (3rd ed.).* Orlando, FL:Harcourt Brace.

Shapiro, S. S. & Wilk, M. B. (1965). An analysis of variance test for normality (complete samples). *Biometrika, 52* (3-4), 591–611.

Smirnov, N. V. (1948). Tables for estimating the goodness of fit of empirical distributions. *Annals of Mathematical Statistics, 19,* 279.

Tabachnick, B. G. & Fidell, L. S. (1996). *Using Multivariate Statistics (3rd ed.).* New York: Harper Collins College Publishers.

van Belle, G. (2002). *Statistical Rules of Thumb.* New York, John Wiley & Sons.

Wolins, L. (1982). *Research mistakes in the social and behavioral sciences.* Ames: Iowa State University Press.

Yates, D., Moore, D., & McCabe, G. (1999). *The Practice of Statistics* (1st Ed.). New York: W.H. Freeman.

Index

A

Adjusted R 217, 329
Analysis of Covariance *see ANCOVA*
Analysis of Variance *see ANOVA*
ANCOVA 192-194, 217-218, 329
ANOVA 35, 131-32, 188-195, 329
Assumptions (Test)
  Binary Logistic Regression 269
  Chi-Square 259
  Correlation 195
  Multiple Linear Regression Analysis 89-144
  One-Way ANOVA 188-189
  t-test (Independent-Samples) 198

B

b see Unstandardized Regression Coefficients
β see Standardized Regression Coefficients
Bivariate Correlation *see Correlation*
Binary Logistic Regression 270-276, 329
Bivariate Analysis 187-203, 254-268, 329
Bivariate Test Key 188, 254
Bonferroni Post Hoc Test 190-192
Boxplot 102-14, 329

C

Categorical Variables 37-38, 184-185, 329
Checks of Data Integrity 85-182, 248-253
Chi-Square Test 259-267, 329
Coding Variables 63-67
Coefficient of Determination 330

X

Y

Z